D0760847

POPULATION BIOLOGY

Springer
New York
Berlin
Heidelberg
Barcelona
Budapest
Hong Kong
London
Milan
Paris
Singapore
Tokyo

Alan Hastings

POPULATION BIOLOGY

Concepts and Models

With 77 Illustrations

Springer

Alan Hastings
Division of Environmental Studies
University of California at Davis
Davis, CA 95616
USA

Original artwork of the flour beetle by Pao Lee Her, who was supported by NSF grants (DMS-9306271) and (DMS-9319073) to R.F. Costantino.

Library of Congress Cataloging-in-Publication Data
Hastings, A. (Alan), 1953–
Population biology : concepts and models / Alan Hastings.
p. cm.
Includes bibliographical references and index.
ISBN 0-387-94862-7 (hardcover : alk. paper).—ISBN 0-387-94853-8 (softcover : alk. paper)
1. Population biology—Mathematical models. I. Title.
QH352.H38 1996
574.5'248'0151—dc20 96-33165

Printed on acid-free paper.

Production managed by Robert Wexler; manufacturing supervised by Joe Quatela.
Camera-ready copy prepared using the author's LaTeX files.
Printed and bound by Hamilton Printing Co., Rensselaer, NY.
Printed in the United States of America.

9 8 7 6 5 4 3 2 (Corrected second printing, 1998)

ISBN 0-387-94862-7 Springer-Verlag New York Berlin Heidelberg SPIN 10547892 (Hardcover)
ISBN 0-387-94853-8 Springer-Verlag New York Berlin Heidelberg SPIN 10675742 (Softcover)

To Elaine, Sara, and Toby

Preface

This book is based on a population ecology course I have taught for many years at the University of California, Davis, primarily to juniors and seniors majoring in Environmental Biology. The material is more than I have ever covered in a ten week quarter, and should be covered comfortably in a semester. This book will also be useful as a supplement to more general texts in population biology and ecology, which tend not to cover nearly enough theory.

The only mathematical prerequisite is one year of calculus (which I assume many students may not remember well). I assume that the student has had some previous exposure to ecological ideas, although this is not an absolute requirement. I believe that the material covered here is essential for any graduate student in ecology or population biology. One principle guiding this book is that students in population biology and ecology are shortchanged if they are not given the opportunity to work through and understand the mathematical models which have become a core part of our discipline. I have not overwhelmed the reader with an excessive number of biological examples in the text – when I teach the course I ask students to read a selection of recent experimental and observational studies to complement the material in the text.

If, after a glance at the table of contents and a look through the book, you begin to doubt that all students in ecology will really be able to do stability analyses, I assure you that students in my classes have had no difficulty rising to the challenge.

I have always tried for the simplest possible explanations of the underlying ideas. For example, I emphasize concepts of age structure in the context of just two age classes. I have also included detailed explanations of all the mathematical concepts and procedures which are likely to be new, at the point they are first used. These are set apart in boxes, so they are easier to find for future reference. Thus, boxes are more frequent in the earlier chapters. Also, since the mathematical development is essentially self-contained, there is more mathematics in the earlier chapters than in the later chapters.

Because the theoretical and mathematical development is such an integral part of the presentation, the book should really be covered in order, with one exception. The chapter on density dependence can be read before the chapter on population genetics. With some care, and review of the boxes in the chapter on population genetics, the chapter on population genetics could be skipped while still maintaining the logical flow of the book.

The extensive marginal notes serve several purposes. Steps that may be confusing for some readers, but clear to others, are explained in the notes. The notes point out connections among different topics. The notes are also used to suggest places where the reader should stop and think (and not just read!).

At the end of every chapter there are problems (some mathematical, some not). Almost all of these have been tested with students. These are an essential part of the text and need to be studied and pondered for a full understanding of the material.

I thank Bill Settle, who took the course many years ago and took careful notes which helped greatly in the preparation of this book. Other students since then have provided extensive feedback on preliminary versions of this text. Emilio Bruna read chapter three. Gary Huxel and Chris Ray have made numerous, helpful comments on the text. Rob Garber, the editor, provided extraordinary help in all phases, including extensive comments on earlier drafts. Finally, I thank Simon Levin, from whom I first learned ecological modeling.

Alan Hastings
Davis, California

Contents

List of Boxes

1
Introduction

1.1 What is population biology?

A *population* is a group of individuals of the same species that have a high probability of interacting with each other. A simple example would be trout in a lake, or moose on Isle Royale, although in many cases the boundaries delineating a population are not as clear cut. Population biology is simply the study of biological populations.

Population biology includes genetic and evolutionary questions; it is broader than population ecology.

Why study population biology? An understanding of complex ecological communities with numerous species interacting with each other and the environment requires an understanding of the simpler ecological systems of one or two species first. We will begin by focusing on the population biology of a single species for two reasons. First, an understanding of the dynamics of a single species leads to the primary questions of population ecology. Second, this is the simplest system that can be studied from a population approach.

Although population ecology might strictly refer to a single species, we use it in this text more broadly to refer to studies focusing on numbers within a species and explicitly consider one or two species at a time.

Why do we focus on numbers of individuals as the variable of interest, rather than a variable like energy flow? We do this because it is possible, even likely, that small numbers of individuals may have effects, especially on population stability, out of

Here 'regulating' means controlling the population growth.

proportion to their numbers. For example, a small popuation of predators may play a major role in regulating a prey population with a large population size. Diseases, ranging from AIDS to the plague, may have extreme effects on the dynamics of the host populations.

Population biology is by its nature a science that focuses on numbers. Thus, we will be interested in understanding, explaining, and predicting changes in the size of populations. Several intriguing patterns of population change through time are illustrated in Figure 1.1. What causes these different patterns to appear? The answer to this question is a central theme of this book.

1.2 Role of models in population biology

Answering many important questions in conservation biology requires the use of models from population biology.

We do not address the important issues of population estimation or experimental design in this book.

The goals of population biology are to understand and predict the dynamics of populations. Understanding, explaining, and predicting dynamics of biological populations will require models, models that are expressed in the language of mathematics. In this book, we emphasize the role of models in understanding population biology. Mathematical models are essential in making precise theoretical arguments about the factors affecting the rate of change of population size.

A full appreciation of the role of models will come as you progress through the book, but a few preliminary observations on how models are employed are very useful. First, a model cannot be shown to be true by a single experiment. However, a model can be shown to be false by a single experiment that does not agree with the predictions of the model. What does it mean to say that a model is false? Assuming that no logical (or mathematical) mistakes have been made, it means that one of the assumptions of the model is not met by the natural system examined. This can be a very useful result, in that it indicates where empirical work should concentrate. In the next chapter we will see just this approach, in elucidating the central question in population biology: *What prevents uncontrolled population growth?*

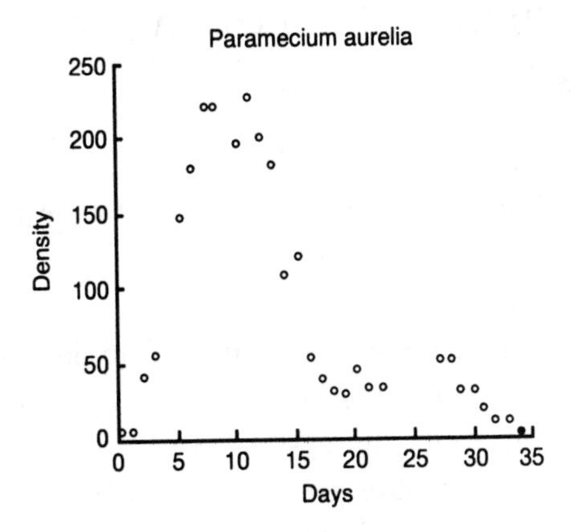

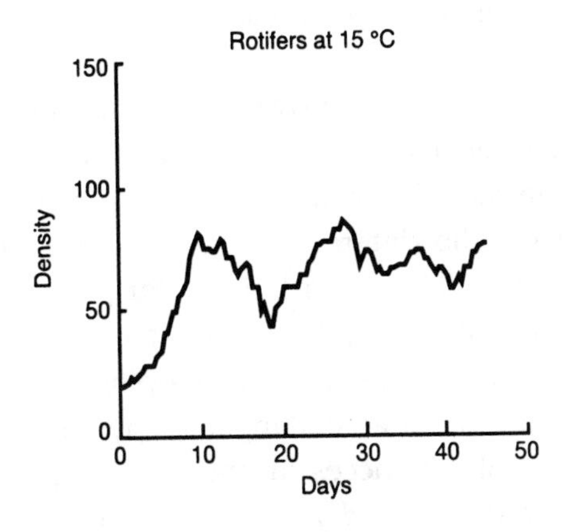

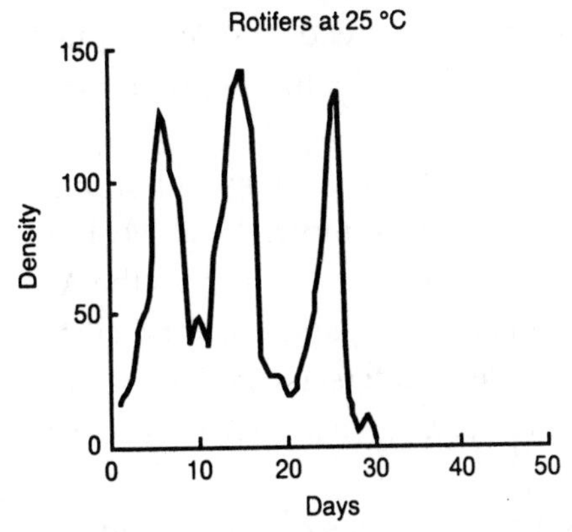

FIGURE 1.1. Three examples of population dynamics from the laboratory. The top example is from Gause (1935); the bottom two are dynamics of rotifers at two different temperatures from Halbach (1979).

1.3 Some successful models in population biology

What are some examples of models that have proven successful in population biology? We provide two brief examples here, one of which we will cover in the text, and one that is beyond the scope of the this book.

One of the most successful theories in population biology has been that of the *dynamics of age–structured population growth*. Given information about the age at which individuals have offspring and the probabilities of death at different ages, we can make detailed predictions about the long term changes in population numbers. We develop the basics of this theory in the next chapter.

Age–structured population growth has played an important role in the field of conservation biology. We can use this theory to ask what changes in survival of different ages, or changes in birth rates, could make the difference between a population that is increasing and one which is declining. Thus, as in studies of the desert tortoise (Doak et al., 1994), this theory can be used to make suggestions that will help in developing a plan to promote the long–term survival of an endangered species.

A second example of successful theory in population biology is the *theory of spatial spread*. As illustrated in Figure 1.2, we can predict the future rate of spread of some populations from initial observations. (Explaining why this is true is beyond the scope of this book.)

Being able to predict the rate of spread of organisms is of great practical importance. For example, the ability to predict the rate of spread of an undesirable species, such as the Africanized honeybee, can help in making preparations for dealing with difficulties caused by this invader. Understanding the rate of spread may also help in the design of control measures to reduce the rate of spread of pests.

These are just two examples of how theory can help to understand questions in population biology. Many more examples of this kind will appear throughout the text.

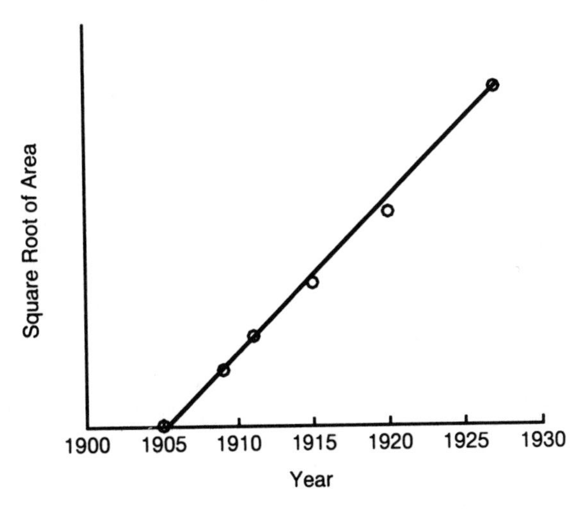

FIGURE 1.2. The rate of spatial spread of the muskrat in central Europe (from Skellam, 1951). As indicated by the straight line (which is fit by eye), the square root of the area occupied by the species increases in an almost perfectly linear manner.

Problems

1. Write a short essay explaining why it is easier to demonstrate that a theory is incorrect than to demonstrate that a theory is correct.
2. Write a short essay explaining the value of theories that can be shown to be incorrect.

Suggestions for further reading

Kingsland (1985) has written a history of population ecology. Virtually any text in ecology will help provide further insight into some of the issues raised here.

A classic paper on the role of models is Levins (1966). See also Pielou (1981). The book *Perspectives in Ecology* by Roughgarden et al. (1989) has a number of essays on the role of theory.

The book *Dynamics of Invasions* by Hengeveld (1989) provides many examples of spatial spread.

Part I

Single species

2
Density-Independent Population Growth

In this chapter, we examine the simplest models of population growth: those which assume density independence. We say that the growth of a population is *density independent* if the birth and death rates per individual do not depend on the population size. We begin with models that also ignore the effects of age structure, and then move on to include the effects of age structure. We look at both species with overlapping generations, like humans, and those with discrete generations, like many butterflies. The predictions of these models set the stage for the rest of our investigations of population dynamics. The approach we use to set up these models will be used again and again. Additionally, we introduce many of the mathematical tools that are used throughout our investigations.

Later, we will look at density dependence.

2.1 The simplest model of population growth

We first examine the simplest model of population growth of a single species. The model begins with two assumptions of density independence:

As with all models, you should consider the importance of factors we ignore. List some factors we have ignored.

These assumptions correspond to density independence because they imply that the per capita birth and death rates are independent of density.

- The rate of births is proportional to the number of individuals present.
- The rate of deaths is proportional to the number of individuals present.

We will look at two models that incorporate these assumptions: one is for populations where the generations overlap and the second for populations in which the generations are nonoverlapping.

Continuous time

We begin with species in which generations overlap and for which births can occur at any time, as in humans. Then it makes sense to predict or count the population size at all times, rather than at specified intervals. Thus, we will use a *continuous time model.* We also need to think about our measure of population size. In this chapter we typically use the individual as the basic unit of population size and count the total number of individuals. However, in some cases it would be both more convenient and realistic to count *biomass*, the total weight of the population. Although counting the number of individuals in the population means the size of the population must be an integer, we do not restrict the population size N at time t to be an integer. Instead, we assume that describing the population by a continuous variable is a reasonable approximation.

Let b be the birth rate per capita and m (for mortality) the death rate per capita. The rate of change of the total number of individuals, $\frac{dN}{dt}$, is given by the rate of births in the population minus the rate of deaths. The rate of births in the population is given by the per capita birth rate times the number of individuals, or bN. Similarly, the population level death rate is given by mN. Thus

$$\frac{dN}{dt} = bN - mN. \tag{2.1}$$

This equation is usually written as

$$\frac{dN}{dt} = rN, \tag{2.2}$$

where $r = b - m$. Note that in this equation the parameter r, the *intrinsic rate of increase*, depends on both the birth rate and the death rate. This model is simple enough that we can solve it completely by finding the population size as a function of time, $N(t)$. We will go through all the steps here because we use a similar process in more complex models. We first separate the variables by putting all the terms with N on one side of the equation and all the terms with t on the other side of the equation:

$$\frac{dN}{N} = rdt. \tag{2.3}$$

Integrate both sides of the equation from $t = 0$ to $t = T$,

Recall that $\int \frac{1}{N} dN = \ln(N)$.

$$\int_{t=0}^{t=T} \frac{dN}{N} = \int_{t=0}^{t=T} rdt. \tag{2.4}$$

Compute the integrals to find

$\ln N(t)|_{t=0}^{t=T}$ means $\ln N(T) - \ln N(0)$.

$$\ln N(t)|_{t=0}^{t=T} = rt|_{t=0}^{t=T}. \tag{2.5}$$

Evaluate the terms in this equation at the specified limits, obtaining

$$\ln(N(T)) - \ln(N(0)) = rT. \tag{2.6}$$

Take the exponential of both sides

$$e^{\ln(N(T))}e^{-\ln(N(0))} = e^{rT}. \tag{2.7}$$

Finally, solve for $N(T)$:

Recall that $e^{\ln(a)} = a$ and $e^{-\ln(a)} = \frac{1}{a}$.

$$\frac{N(T)}{N(0)} = e^{rT} \tag{2.8}$$

$$N(T) = N(0)e^{rT}. \tag{2.9}$$

This model predicts that

- if $r = 0$, the population size is stationary
- if $r > 0$, the population grows exponentially without bound
- if $r < 0$, the population approaches 0.

These dynamics are illustrated in Figure 2.1.

Solve Problem 1.

There are important biological and mathematical conclusions we wish to draw from this simple model. First, the model can

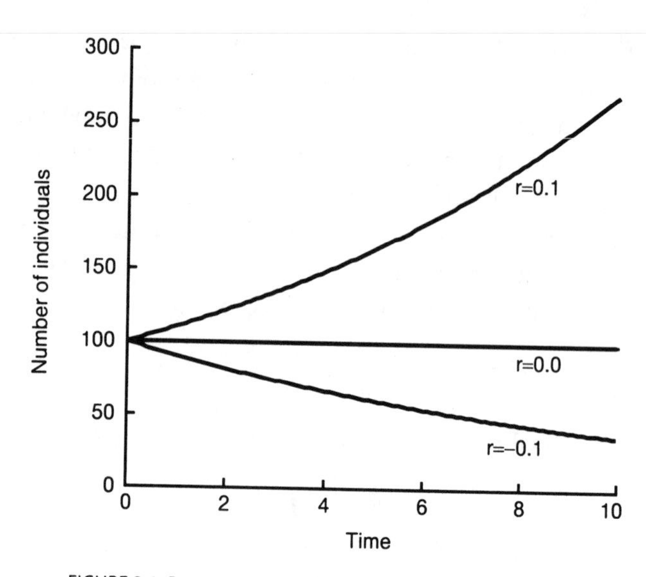

FIGURE 2.1. Dynamics of continuous time population growth.

help answer the question: do biological populations ever demonstrate exponential growth? Yes, but only for short times, because exponential growth is so rapid that populations cannot grow exponentially for long. Second, an equilibrium population is only attained when r is exactly 0, which is clearly an unlikely event. Consequently, populations that remain constant over a prolonged period of time, or even approximately constant, cannot be explained by this model. Thus, the density-independent model cannot explain most populations we see in nature.

The model cannot tell us why a population does not keep growing exponentially, but perhaps you have some ideas.

Third, we will use the exponential solution to the linear growth equation (2.2) many times. Commit to memory the relationship between the sign of r and the fate of the population, as this qualitative conclusion is very important. We emphasize this in the box.

Equation (2.2) is called 'linear' because the right-hand side only has terms that contain N; there are no constant terms and no terms with N raised to a power. Linear equations are special, as we shall see throughout the book.

Discrete time

Is the continuous time model a good choice for all species? Recall our assumption about overlapping generations and births occurring at any time. Butterflies like the bay checkerspot *Euphydrias editha* breed once per year, laying their eggs close to April 1. Adults only fly for a short period, and then die. Large ungulates,

Box 2.1. Solution for exponential growth in continuous time.

Here we summarize equation (2.3) - (2.9). The solution to

$$\frac{dN}{dt} = rN$$

is

$$N(t) = N(0)e^{rt}.$$

This equation and its solution will be used many times. The *qualitative* behavior of the solution is determined only by the sign of r.

- For $r > 0$, the solution increases without bound,
- for $r = 0$, the solution is a constant, and
- for $r < 0$ the solution approaches zero.

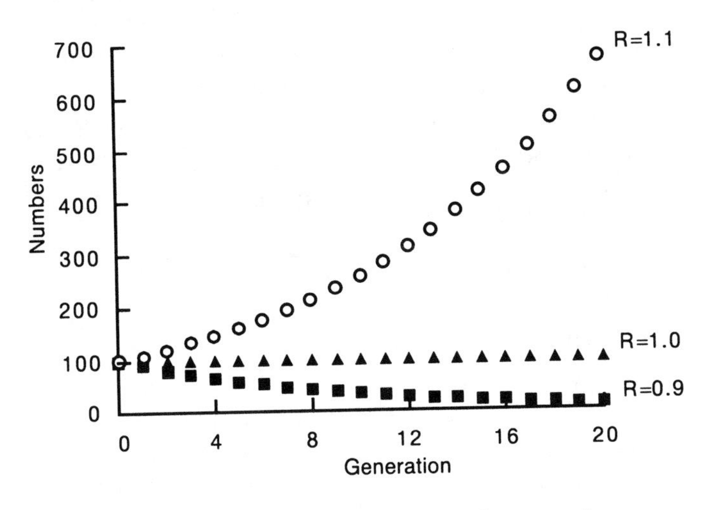

FIGURE 2.2. Dynamics of discrete time population growth.

such as moose, have offspring only once per year in the spring, with adults rarely living more then 10 years. For these species, a model which assumes both that births occur continuously and that generations overlap is inappropriate.

Think of other species for which the continuous model is inappropriate.

Which model do you think would be best for humans?

For univoltine (one generation per year) insects like the bay checkerspot or annual plants, a *discrete time* model, with popu-

Box 2.2. Solution for exponential growth in discrete time.

The solution to

$$N_{t+1} = RN_t$$

is

$$N_t = R^t N_0.$$

This equation and its solution will be used many times as well. The *qualitative* behavior of the solution is determined only by the difference between R and 1.

- For $R > 1$, the solution increases without bound
- for $R = 1$, the solution is a constant
- for $R < 1$, the solution approaches 0.

lation measurements only taken at fixed times, is more appropriate. We will measure time in units of generations, which may be 1 year. Here we will let R be the number of individuals in the next generation per individual in the current generation. Thus, if N_t is the number of individuals in the population at generation t,

We will see the contrast between discrete time and continuous time arising throughout the book. We would expect that in most cases the conclusions we draw should not depend on this difference. However, in several cases the difference between discrete time and continuous time is critical.

$$N_{t+1} = RN_t. \tag{2.10}$$

From the relationship

$$N_2 = RN_1 = R(RN_0) = R^2 N_0 \tag{2.11}$$

Use $N_1 = RN_0$ in equation (2.11).

we conclude that

$$N_t = R^t N_0. \tag{2.12}$$

After 2 years there is a 2 in the exponent so we expect that after t years there will be a t in the exponent.

Thus, once again, we see that growth is exponential or geometric. The qualitative behavior of the solution is illustrated in Figure 2.2 and discussed in Box 2.2. Note that we reach the same biological and mathematical conclusions from this model that we reached using the continuous time model.

Relationship between continuous and discrete models

We have two models differing only in the assumption of discrete versus overlapping generations. What is the connection between these two models? Comparing the solutions of the two models given in equations (2.9) and (2.12), we see that R^t and $e^{rt} = (e^r)^t$ play analogous roles. Thus we conclude that

$$R = e^r \tag{2.13}$$

or

$$\ln(R) = r. \tag{2.14}$$

If r is small, or analogously if R is close to 1, we can use Taylor series to find the approximate relationship:

The Taylor series for e^r is $e^r \approx 1 + r + \frac{r^2}{2} + \cdots$.

$$R \approx 1 + r. \tag{2.15}$$

Can you think of a heuristic explanation why this equation has a 1 in it and is not simply $R \approx r$?

In continuous time, $r = 0$ corresponds to a population that remains the same size, while in discrete time $R = 1$ corresponds to a population that remains the same size.

We conclude that both the continuous time and discrete time models make similar predictions. As we consider more realistic situations we will extend both these models, often choosing which one to use on the basis of mathematical convenience as well as for biological reasons.

Exponential growth in nature

We have seen in Figures 2.1 and 2.2 some hypothetical examples of population growth. What are typical values of per capita growth rates in laboratory and natural populations? In the laboratory, estimates of r or R can show quite rapid population growth, but in natural populations r is almost always near zero and R near one. As we have noted, a population that does not show such values of growth will either explode in numbers or disappear rapidly.

One example of exponential growth in nature is illustrated in Figure 2.3. Here the collared dove numbers initially grew approximately exponentially, but after some years, the growth was no longer exponential. In Chapter 4, we will try to understand this aspect of population dynamics.

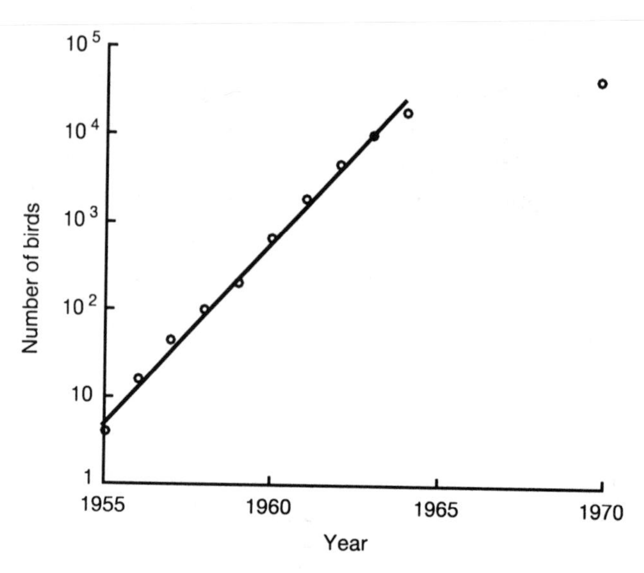

FIGURE 2.3. Dynamics of population growth for collared dove in Great Britain. Note the log scale, so the straight line fit through the population counts implies the growth is approximately exponential until about 1963 (data from Hudson, 1965, 1972). The last data point (1970), which is far below the level that would be reached if exponential growth continued, is an estimate because the population had become too large to count accurately.

The fundamental question of population ecology

Population sizes reach absurdly high levels if exponential growth continues for more than a short time. Here, we admit some circularity – we are calling a time short if it is short enough that exponential growth makes sense. Problem 1 helps make this idea more concrete.

What can we conclude about the simple models that we have just considered? As noted, the prediction of exponential growth can be valid only for a short period of time. Thus the assumptions of the models, in particular the assumption that the rate of population growth is proportional to the number of individuals (the assumption of *density independence*), must be in error. The fundamental question of population biology is to determine the causes and consequences of the deviation from exponential growth, or simply what regulates populations.

2.2 Age structure: the simplest case

Before we examine the consequences of density dependence, we continue our study of density independence by looking at models more appropriate for organisms that reproduce many times, still assuming density-independent birth and death rates.

Many species of interest to ecologists do not have the simple life history that allows us to blindly use the models we have just developed. The common snapping turtle (*Chelydra serpentina*) has a life history typical of many other turtles (Congdon et al., 1994). Females do not become sexually mature until they are more than 5 years old, and continue to lay eggs essentially throughout their life, which may be as long as 100 years. How should we understand the dynamics of this species? The models we have developed do not seem to apply. Understanding the dynamics of long-lived species is important for many conservation questions, as many long-lived species are endangered.

A life history is the pattern of births and deaths (and possibly dispersal patterns) for a population.

Many mammals also are not well described by the simple models we have just developed. Humans, for example, have different fecundities as a function of age, as do virtually all but the simplest animals or plants. If, in addition, generations overlap so that different ages are present at the same time, we need to modify our models because we have thus far assumed that generations do not overlap.

In general, mammals do exhibit one characteristic that allows us to simplify our models. We will focus on females, looking only at the reproduction of females and only at female births. This assumption is justified if males are not limiting reproduction in the species.

Can you think of populations where the simplification of considering only females would be wrong? Why?

A very simple model

Before considering forces that prevent populations from growing exponentially, we will consider whether our prediction of exponential growth holds in a slightly more complex setting where we include age structure. Rather than start with a general model, we will study a hypothetical organism that lives for 2 years, potentially reproducing either at age 1 or at age 2.

Most biennial plants, plants that reproduce after 2 years, are unable to reproduce after 1 year.

We assume, as earlier, that the per capita birth rate and the survival rate are unaffected by the number of individuals present. However, birth rates and death rates will depend on the ages of the individuals. Count individuals at the end of any season, and

The approach we take, of looking at an idealized – perhaps unrealistically simplified – case that does however capture the essence of a biological interaction, is a very important approach for elucidating general principles.

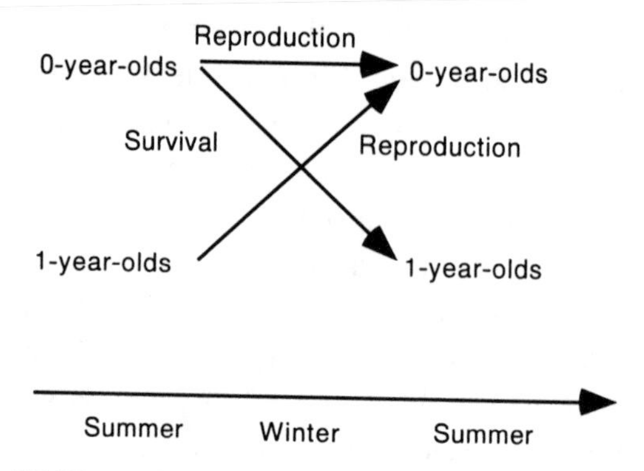

FIGURE 2.4. Life history of a population with two age classes

There is nothing special about 83 and 82; we are choosing two years to use as examples.

determine the growth rate. For example, look at the end of the 83rd season to get $R = N_{83}/N_{82}$. Without age structure, we would have exponential growth with $N_{84} = RN_{83}$. Does this relationship hold even with age structure?

In answering this question we make a fundamental assumption by considering only females and only female births in populations with two sexes, as we noted. Next, we need to set up the parameters in a model to answer this question. We will call an organism 0 years old during its first year of life and 1 year old during its second year. We will assume that all individuals die before they reach their third year. The parameters and variables we need to describe this are as follow:

- m_0 is the mean number of offspring of a 0-year-old the following year.
- m_1 is the mean number of offspring of a 1-year-old the following year.
- S_0 is the probability that a 0-year-old survives to become a 1-year-old.
- $n_0(t)$ is the number of 0-year-olds at time t.
- $n_1(t)$ is the number of 1-year-olds at time t.
- $N(t) = n_0(t)+n_1(t)$ is the total number of organisms at time t.

We first describe the model in words, and then in equations.

$$\begin{aligned} n_0(t+1) = &\text{ number of offspring of 0-year-olds in year } t \\ &+ \text{number of offspring of 1-year-olds in year } t \end{aligned}$$

$n_1(t+1)$ = number of 0-year-olds in year t times the probability of survival from 0 to 1

Translating this into equations, we find that

$$n_0(t+1) = n_0(t)m_0 + n_1(t)m_1 \tag{2.16}$$

$$n_1(t+1) = n_0(t)S_0. \tag{2.17}$$

We are now prepared to answer the question we just posed: does a population grow exponentially even if age structure in reproduction is important? We can rephrase this question as: is there a unique growth rate for the population that is not dependent on how much time has elapsed since the population was established? What happens if in year 0 we have 10 zero-year-olds and no one-year-olds, so that $n_0(0) = 10$ and $n_1(0) = 0$ are the starting sizes in each age class? Using equations (2.16) and (2.17), we see that next year

$$n_0(1) = 10m_0 \tag{2.18}$$

$$n_1(1) = 10S_0. \tag{2.19}$$

On the other hand, if in year 0 we have no zero-year-olds and 10 one-year-olds, so that $n_0(0) = 0$ and $n_1(0) = 10$, using (2.16) and (2.17), we see that next year

$$n_0(1) = 10m_1 \tag{2.20}$$

$$n_1(1) = 0. \tag{2.21}$$

These are very different populations (unless of course we choose special values for the parameters), so we conclude that no single number represents a growth rate.

For example, let $m_0=1$, $S_0 = \frac{1}{2}$, and $m_1 = 2$. Then, starting with all zero-year-olds, we have 10 zero-year-olds and 5 one-year-olds in the first year. However, starting with all one-year-olds, we have 20 zero-year-olds and no one-year-olds in the first year.

How could we get a specific, constant growth rate? Every 0-year-old always makes the same contribution to the population the next year, but the contribution of 1-year-olds is different from that of 0-year-olds. If the ratio of 0-year-olds to 1-year-olds stays the same, then we expect that the growth rate of the population stays the same.

Are there conditions where the ratio of 0-year-olds to 1-year-olds remains the same? Let us assume that there are and see what happens. We will see later that the assumption that the growth rate stays the same implies that the ratio of 0-year-olds to 1-year-olds stays the same and vice versa. First let us assume that $n_1(t) = cn_0(t)$, where c is the ratio of one-year-olds to zero-year-olds. Then, substituting into the model (2.16) and (2.17), we find

To get these equations, we replace $n_1(t)$ everywhere it appears in (2.16) by $cn_0(t)$.

$$n_0(t+1) = n_0(t)m_0 + cn_0(t)m_1 \tag{2.22}$$

$$cn_0(t+1) = n_0(t)S_0. \tag{2.23}$$

In (2.22) and (2.23), we view the ratio c as the unknown.

We now try to find a ratio c that is constant from generation to generation, by eliminating $n_0(t+1)$ and $n_0(t)$ from equations (2.22) and (2.23). From (2.23) we find

$$n_0(t+1) = n_0(t)S_0/c. \tag{2.24}$$

Substituting this expression into (2.22) we get

$$n_0(t)S_0/c = n_0(t)m_0 + cn_0(t)m_1. \tag{2.25}$$

Dividing by $n_0(t)$ yields a quadratic equation for c, which we can solve using the quadratic formula:

$$S_0/c = m_0 + cm_1 \tag{2.26}$$

$$S_0 = m_0c + c^2m_1 \tag{2.27}$$

$$0 = m_1c^2 + m_0c - S_0 \tag{2.28}$$

$$c = \frac{-m_0 \pm \sqrt{m_0^2 + 4S_0m_1}}{2m_1} \tag{2.29}$$

Thus there are two possible solutions c for a ratio of 0-year-olds to 1-year-olds that remains constant from year to year. If the sign is + in (2.29), the ratio c is positive, while if the sign is − the ratio is negative. From (2.24) we see that the growth rate of the population is S_0/c, so for the positive value of c, S_0/c is a candidate for the growth rate of the population. We will return to the meaning of the negative possibility.

The growth rate is $n_0(t+1)/n_0(t)$.

What happens if the population does not start off at the 'magic' ratio of 0-year-olds to 1-year-olds? As we will see below, through time the population will approach both the ratio of 0-year-olds and 1-year-olds and the growth rate we have just found.

Box 2.3. Matrices and vectors.

We will find matrices and vectors useful for expressing many of the models we study, as a compact notation for expressing linear equations.

- A *matrix* is a rectangular array of numbers, for example

$$\begin{pmatrix} a_{11} & a_{12} \\ a_{21} & a_{22} \end{pmatrix}.$$

  This is a 2×2 matrix, where the first 2 refers to the number of rows and the second 2 refers to the number of columns.
- A column *vector* is a matrix with only 1 column

$$\begin{pmatrix} u_1 \\ u_2 \end{pmatrix}.$$

- A row vector is a matrix with only one row.

$$\begin{pmatrix} u_1 & u_2 \end{pmatrix}.$$

When we say vector in this text we almost always mean column vector.

Using matrices to write the model

A much more convenient notation, essential for extending the model to include more age classes, is to use *matrices*. If you are unfamiliar with matrix notation and matrix multiplication, these concepts are described in Boxes 2.3 through 2.7. We can write the model (2.16) and (2.17) using matrices as

$$\begin{pmatrix} m_0 & m_1 \\ S_0 & 0 \end{pmatrix} \begin{pmatrix} n_0(t) \\ n_1(t) \end{pmatrix} = \begin{pmatrix} n_0(t+1) \\ n_1(t+1) \end{pmatrix}. \quad (2.30)$$

As we have just done, we will assume that the ratio of 0-year-olds to 1-year-olds remains constant and the population grows at the rate λ per year. (In the previous section, the growth rate was S_0/c.) We call this ratio of individuals at different ages that remains constant a *stable age distribution*. As we will see, not only does the ratio of the numbers of individuals at different ages remain

We will see the concept of stability throughout this book.

Box 2.4. Matrix addition and subtraction.

The general rule for adding or subtracting matrices of vectors is that corresponding elements are added or subtracted. Thus,

$$\begin{pmatrix} a_{11} & a_{12} \\ a_{21} & a_{22} \end{pmatrix} + \begin{pmatrix} b_{11} & b_{12} \\ b_{21} & b_{22} \end{pmatrix} = \begin{pmatrix} a_{11}+b_{11} & a_{12}+b_{12} \\ a_{21}+b_{21} & a_{22}+b_{22} \end{pmatrix}.$$

As an example:

$$\begin{pmatrix} 2 & 3 \\ 4 & 5 \end{pmatrix} - \begin{pmatrix} 1 & 2 \\ 3 & 1 \end{pmatrix} = \begin{pmatrix} 1 & 1 \\ 1 & 4 \end{pmatrix}.$$

Box 2.5. Matrix multiplication.

The general formula for multiplying a matrix times a vector is:

$$\begin{pmatrix} a_{11} & a_{12} \\ a_{21} & a_{22} \end{pmatrix} \begin{pmatrix} u_1 \\ u_2 \end{pmatrix} = \begin{pmatrix} a_{11}u_1 + a_{12}u_2 \\ a_{21}u_1 + a_{22}u_2 \end{pmatrix}$$

As an example:

$$\begin{pmatrix} 2 & 3 \\ 4 & 5 \end{pmatrix} \begin{pmatrix} 6 \\ 7 \end{pmatrix} = \begin{pmatrix} (2 \cdot 6) + (3 \cdot 7) \\ (4 \cdot 6) + (5 \cdot 7) \end{pmatrix} = \begin{pmatrix} 33 \\ 59 \end{pmatrix}.$$

constant at the stable age distribution, but *the stable age distribution is approached from other age distributions*. This is why we use the word stable, to emphasize that this age distribution is approached. The notion of a stable age distribution is central to the understanding of the dynamics of age-structured populations.

One fact about the stable age distribution is important to emphasize here. The stable age distribution refers to the ratios of the numbers of individuals at different ages and not to the absolute numbers. Thus, the stable age distribution is not unique – if 10 zero-year-olds and 5 one-year-olds is a stable age distribution, so is 20 zero-year-olds and 10 one-year-olds.

Box 2.6. Multiplying a matrix by a scalar.

A scalar is just a single number. To multiply a matrix by a scalar, simply multiply each element in the matrix by the scalar.

$$\lambda \begin{pmatrix} a_{11} & a_{12} \\ a_{21} & a_{22} \end{pmatrix} = \begin{pmatrix} \lambda a_{11} & \lambda a_{12} \\ \lambda a_{21} & \lambda a_{22} \end{pmatrix}.$$

As an example:

$$4 \begin{pmatrix} 2 & 3 \\ 4 & 5 \end{pmatrix} = \begin{pmatrix} 8 & 12 \\ 16 & 20 \end{pmatrix}$$

We now express the assumption that the population is in a stable age distribution in matrix and vector notation. The idea is that in every year the ratio of 0-year-olds to 1-year-olds remains constant, but that the number in each age class grows at the rate λ each year. In vector and matrix notation, we write

There is no t in (2.31) because we assume that the stable age equation holds year after year.

$$\begin{pmatrix} m_0 & m_1 \\ S_0 & 0 \end{pmatrix} \begin{pmatrix} n_0 \\ n_1 \end{pmatrix} = \begin{pmatrix} \lambda n_0 \\ \lambda n_1 \end{pmatrix}. \tag{2.31}$$

In honor of P.H. Leslie, who was one of the first to describe the use of matrices to describe population dynamics with age structure in papers published in 1945 and 1948, we will call the matrix in this equation a *Leslie matrix* and denote it by L. We will let N be the vector of population sizes, so

$$N = \begin{pmatrix} n_0 \\ n_1 \end{pmatrix} \tag{2.32}$$

We can then write our equation for the stable age distribution in the even more compact form

$$LN = \lambda N \tag{2.33}$$

What is noteworthy about this expression is that it is valid for an arbitrary number of age classes. We wish to determine a formula for λ from this expression, but we cannot simply divide by the vector N. Beginning with (2.33) we write

Based on our example in the previous section we expect two possible solutions for λ.

Box 2.7. The identity matrix.

If we multiply any number by 1 the answer is just the number we started with. The number 1 is called the (multiplicative) identity. The matrix that plays the same role is called the identity matrix and is denoted by I, so $IN = N$ for any matrix or vector N of the appropriate size. The identity matrix with two rows and columns is

$$I = \begin{pmatrix} 1 & 0 \\ 0 & 1 \end{pmatrix}.$$

Try it and see.

$$LN - \lambda N = 0 \tag{2.34}$$

$$LN - \lambda IN = 0 \tag{2.35}$$

$$(L - \lambda I)N = 0 \tag{2.36}$$

where I is the identity matrix, as explained in Box 2.7. A fact from matrix algebra is that the final equation is satisfied (for nonzero N) if and only if the *determinant* of $(L - \lambda I)$ is 0. We will use this fact without trying to justify it. The formula for the determinant of a 2×2 matrix is

We want N to be nonzero as this is a vector of population sizes.

$$\det \begin{pmatrix} a_{11} - \lambda & a_{12} \\ a_{21} & a_{22} - \lambda \end{pmatrix} = (a_{11} - \lambda)(a_{22} - \lambda) - a_{12}a_{21}. \tag{2.37}$$

Setting this determinant to be 0 yields a quadratic equation for the growth rate λ.

We will work through an example. Let $m_0 = 1$, $m_1 = 4$, and $S_0 = 1/2$. Then,

$$L = \begin{pmatrix} 1 & 4 \\ 1/2 & 0 \end{pmatrix} \tag{2.38}$$

and

$$L - \lambda I = \begin{pmatrix} 1 - \lambda & 4 \\ 1/2 & -\lambda \end{pmatrix}. \tag{2.39}$$

Setting the determinant of $L - \lambda I$ equal to 0 yields

Here the equation satisfied by λ, (2.41), can be factored, but in general we have to use the quadratic formula.

$$(1 - \lambda)(-\lambda) - 2 = 0 \tag{2.40}$$

$$\lambda^2 - \lambda - 2 = 0 \quad (2.41)$$

$$(\lambda - 2)(\lambda + 1) = 0, \quad (2.42)$$

so the two solutions are $\lambda = 2$ or $\lambda = -1$. We will substitute each growth rate in turn into (2.33), with L given by (2.38), to find the stable age distribution we defined earlier and to understand what the second 'growth rate' really means. With $\lambda = 2$, we find that (2.33) is

$$\begin{pmatrix} 1 & 4 \\ 1/2 & 0 \end{pmatrix} \begin{pmatrix} n_0 \\ n_1 \end{pmatrix} = \begin{pmatrix} 2n_0 \\ 2n_1 \end{pmatrix} \quad (2.43)$$

Both the linear equations expressed by this single matrix equation are satisfied if $n_0 = 4n_1$, which is the stable age distribution. Thus if there are four times as many 0-year-olds as there are 1-year-olds, the population will double each year.

The first equation is $n_0 + 4n_1 = 2n_0$ which simplifies to $n_0 = 4n_1$.

If we choose $\lambda = -1$, we find that (2.33), with L given by (2.38), is

$$\begin{pmatrix} 1 & 4 \\ 1/2 & 0 \end{pmatrix} \begin{pmatrix} n_0 \\ n_1 \end{pmatrix} = \begin{pmatrix} -n_0 \\ -n_1 \end{pmatrix} \quad (2.44)$$

Both the linear equations expressed by this single matrix equation are satisfied if $n_0/2 = -n_1$. Does this second 'growth rate' have any biological meaning? A numerical example will help answer this question.

What happens if the population starts with 6 zero-year-olds and no one-year-olds? We compute the dynamics as follows:

We write our starting vector of population size as the sum of the two vectors we just found.

$$\begin{pmatrix} 1 & 4 \\ 1/2 & 0 \end{pmatrix} \begin{pmatrix} 4+2 \\ 1+(-1) \end{pmatrix} = \begin{pmatrix} 8+(-2) \\ 2+1 \end{pmatrix} = \begin{pmatrix} 6 \\ 3 \end{pmatrix} \quad (2.45)$$

$$\begin{pmatrix} 1 & 4 \\ 1/2 & 0 \end{pmatrix} \begin{pmatrix} 8+(-2) \\ 2+1 \end{pmatrix} = \begin{pmatrix} 16+2 \\ 4+(-1) \end{pmatrix} = \begin{pmatrix} 18 \\ 3 \end{pmatrix} \quad (2.46)$$

$$\begin{pmatrix} 1 & 4 \\ 1/2 & 0 \end{pmatrix} \begin{pmatrix} 16+2 \\ 4+(-1) \end{pmatrix} = \begin{pmatrix} 32+(-2) \\ 8+1 \end{pmatrix} = \begin{pmatrix} 30 \\ 9 \end{pmatrix} \quad (2.47)$$

Do you notice a pattern? From the first time step we can 'break up' the numbers in the population into two components, one corresponding to the stable age distribution, and the other corresponding to a deviation from the stable age distribution. With increasing time, the deviation from the stable age distribution rep-

TABLE 2.1. Dynamics with age structure: a numerical example.

year	0-year-olds	1-year-olds	total	growth rate	ratio
0	6	0	6		
1	6	3	9	1.500	2.000
2	18	3	21	2.333	6.000
3	30	9	39	1.857	3.333
4	66	15	81	2.077	4.400
5	126	33	159	1.963	3.818
6	258	63	321	2.019	4.095
7	510	129	639	1.991	3.953
8	1026	255	1281	2.005	4.024
9	2046	513	2559	1.998	3.988
10	4098	1023	5121	2.001	4.006
11	8190	2049	10239	1.999	3.997
12	16386	4095	20481	2.000	4.001

resents a smaller and smaller fraction of the total population. *We have shown that the stable age distribution is indeed stable – it is approached from other age distributions.* Because the stable age distribution is approached through time, an observation of a population that is not in a stable age distribution is evidence that the population has been disturbed at some time.

Can we write down, in a simple form, what the population will be after t years? As in the model without age structure, we start with the change in population over 1 year, $N_1 = LN_0$, and deduce the long-term description,

This is the analog to equation (2.11).

$$N_t = L^t N_0, \tag{2.48}$$

where L^t means to multiply by the Leslie matrix t times. To understand the long-term dynamics, we need a simple expression for $L^t N_0$.

We can express the population after an arbitrary number of years, $L^t N_0$, in a simple form using *eigenvectors* and *eigenvalues* (defined in Boxes 2.8 and 2.9). Let λ_0 and λ_1 be the two eigenvalues and v_0 and v_1 be the two corresponding eigenvectors. Our first step is to express the initial population size as the sum of the

Eigenvalues will also play an essential role in our study of interactions between two species.

two eigenvectors, so

$$N_0 = a_0 v_0 + a_1 v_1, \tag{2.49}$$

where the constants a_0 and a_1 can be determined from solving linear equations.

Using this expression for the initial population, we can find the population after t years as

$$N_t = L^t N_0 = L^t(a_0 v_0 + a_1 v_1) \tag{2.50}$$

$$= a_0 L^t v_0 + a_1 L^t v_1 \tag{2.51}$$

$$= a_0 \lambda_0^t v_0 + a_1 \lambda_1^t v_1. \tag{2.52}$$

To go from (2.50) to (2.51) use the distributive law, and remember that a_0 is a scalar (a single number), so we can move it to the left of L^t.

To go from (2.51) to (2.52), remember that $Lv_0 = \lambda_0 v_0$. Do this t times for each eigenvector.

From equation (2.52), we can read off the important property of populations approaching the stable age distribution. Assume that we have labeled the eigenvalues so that $|\lambda_0| > |\lambda_1|$; then λ_0 is what we have been calling the *growth rate* of the population. After a long time, as t becomes large, the term $a_0 \lambda_0^t v_0$ in (2.52), the expression for the population size after an arbitrary number of years, will be much larger than the term $a_1 \lambda_1^t v_1$. In fact, ignoring the smaller term, we will have $N_t \approx a_0 \lambda_0^t v_0$, so the population will be in the stable age distribution.

What role will the relative sizes of the eigenvalues play in the rate of approach to the stable age distribution? The answer again lies in (2.52).

In this chapter, we have focused for simplicity on two age classes. For an arbitrary number of age classes, our conclusions about the approach to a stable age distribution still hold so long as each age class has a possibility of a descendant in any given year far enough in the future.

Iteroparity versus semelparity

An organism that reproduces just once during its lifetime is called *semelparous*, while an organism that reproduces many times is called *iteroparous*. Examples of semelparous organisms are salmon and periodical cicadas. Oak trees, humans, and turtles are examples of iteroparous species. For organisms that live only 2 years, what would the Leslie matrix of a semelparous organism look like? There would be a zero in the upper left hand corner:

$$\begin{pmatrix} 0 & m_1 \\ S_0 & 0 \end{pmatrix} \tag{2.53}$$

Box 2.8. Eigenvalues of a 2×2 matrix.

We will refer to the formula for the eigenvalues (growth rates) many times. For the 2×2 matrix

$$\begin{pmatrix} a_{11} & a_{12} \\ a_{21} & a_{22} \end{pmatrix}$$

the two eigenvalues λ are the two solutions of the quadratic equation

$$\begin{aligned} 0 &= (a_{11} - \lambda)(a_{22} - \lambda) - a_{12}a_{21} \\ &= \lambda^2 - (a_{11} + a_{22})\lambda + (a_{11}a_{22} - a_{12}a_{21}). \end{aligned}$$

The two eigenvalues can be found by solving this equation using the quadratic formula.

Note that in this case the number of 0-year-olds in year 1 has no effect on the number of 0-year-olds in year 2; there really are two separate populations corresponding to the two year classes. In this case, there is no approach to the stable age distribution. Populations of species that come close to this idealized situation of reproducing only once at a fixed age approach the stable age distribution very slowly.

Data

How do we obtain the data used to estimate population growth rates using the model we have just developed? We will consider the more general case where there is an arbitrary number of age classes, noting only that the length of each age class must be the same. We call the description of age-specific birth and death rates the *life table*. The quantities that we will use in the life table are as follow:

Why is $l_0 = 1$?

- l_x is the fraction of newborn individuals that survive to age x.

What is the relationship among l_x, l_{x+1}, S_x?

- S_x is the probability of surviving from age x to age $x + 1$.
- m_x is the average birth rate for an individual of age x.

Box 2.9. Eigenvectors of a 2×2 matrix.

We will need to use the formula for eigenvectors many times. For the 2×2 matrix

$$L = \begin{pmatrix} a_{11} & a_{12} \\ a_{21} & a_{22} \end{pmatrix}$$

there are two eigenvalues, λ_0 and λ_1. To find the eigenvector v_0 corresponding to the eigenvalue λ_0 that satisfies $Lv_0 = \lambda_0 v_0$ first try a vector of the form

$$\begin{pmatrix} 1 \\ b \end{pmatrix}$$

which leads to the equations

$$a_{11} + a_{12}b = \lambda_0$$
$$a_{21} + a_{22}b = \lambda_0 b$$

for b. In general each equation will have the same solution for b. If these equations do not have a solution for b, try a vector of the form

$$\begin{pmatrix} b \\ 1 \end{pmatrix},$$

which leads instead to the equations

$$a_{11}b + a_{12} = \lambda_0 b$$
$$a_{21}b + a_{22} = \lambda_0$$

for b.
The second eigenvector is found in a similar fashion.

There are two different approaches for obtaining the data in a life table, with different advantages and disadvantages.

Vertical life tables

For organisms that live only a short time (less than the length of time spent collecting the data), one can determine the schedule of age-dependent per capita birth and death rates by following a sin-

Box 2.10. Finding the stable age distribution and growth rate of a population with two age classes.

We will give an example of the whole procedure. Let the birth rate of 0-year-olds be 2, the birth rate of 1-year-olds 4, and the survival rate 1/4. Then the Leslie matrix is

$$\begin{pmatrix} 2 & 4 \\ 1/4 & 0 \end{pmatrix}.$$

The two growth rates are the solutions of the quadratic equation

$$\lambda^2 - 2\lambda - 1 = 0.$$

Using the quadratic formula we find the solutions

$$\lambda_0, \lambda_1 = \frac{2 \pm \sqrt{8}}{2}.$$

Thus, the larger value of λ corresponding to the growth rate is 2.414. If we look for a stable age distribution of the form $Lv_0 = \lambda_0 v_0$, we first try a vector of the form

$$\begin{pmatrix} 1 \\ b \end{pmatrix}.$$

We find that b satisfies

$$2 + 4b = 2.414$$

so a stable age distribution is

$$\begin{pmatrix} 1 \\ 0.1035 \end{pmatrix}.$$

Remember that the stable age distribution is not unique, so that

$$\begin{pmatrix} 4 \\ 0.414 \end{pmatrix}$$

is also a stable age distribution.

TABLE 2.2. Life table for the vole *Microtus agrestis* reared in the laboratory (data from Leslie and Ranson 1940).

age (weeks)	x	l_x	m_x
0	0	1.0000	0.0000
8	1	0.8335	0.6504
16	2	0.7313	2.3939
24	3	0.5881	2.9727
32	4	0.4334	2.4662
40	5	0.2928	1.7043
48	6	0.1813	1.0815
56	7	0.1029	0.6683
64	8	0.0535	0.4286
72	9	0.0255	0.3000

gle *cohort* through time. Start with a fixed number of organisms, and record births and deaths through time. A major disadvantage of this method is the requirement that the maximum age of the organism investigated must be short enough. One example of data of this kind is in Table 2.2. Many more examples are included in Andrewartha and Birch (1954). This is called a *vertical life table* because we think of time as running down the page on which the observations are written.

A cohort is all the individuals born at a single time.

Horizontal life tables

For very long-lived organisms, such as elephants, turtles, and most trees, following a single cohort through time is impractical. Instead, one collects data from a single point in time and estimates birth rates and death rates. This method has several disadvantges because of the implicit assumptions made. In particular, as we will soon see, the ratio of 5-year-olds to 4-year-olds in the current population is usually not the probability of survival from 4 to 5 years. There also needs to be a way of accurately estimating the age of organisms.

This way of getting the life history parameters is called a *horizontal life table*, because the observations are made at a single time. In fact, this approach usually involves estimating some of

TABLE 2.3. Life table for the snapping turtle (Congdon et al., 1994). For $x > 14$, $l_x = 0.93 l_{x-1}$, which is an estimate.

age in years (x)	S_x	l_x	m_x
0	0.230	1.00000	0.0
1	0.470	0.23000	0.0
2	0.810	0.10810	0.0
3	0.650	0.08756	0.0
4	0.650	0.05691	0.0
5	0.750	0.03699	0.0
6	0.740	0.02775	0.0
7	0.810	0.02053	0.0
8	0.770	0.01663	0.0
9	0.800	0.01281	0.0
10	0.820	0.01024	0.0
11	0.820	0.00840	4.0
12	0.820	0.00689	6.0
13	0.930	0.00565	8.0
14	0.930	0.00525	10.0
15 – 105	0.930		12.0

the life table parameters. This is illustrated by the example given in Table 2.3.

With detailed data on age-dependent births and deaths such as those for the snapping turtle or vole, one could construct a large Leslie matrix. However, this is an unwieldy approach. There is an alternative approach that is easier to apply in practice. We will first phrase the approach as a *continuous time model*, and then explain how to use data collected in discrete time, such as the life table data we have just seen, in the model.

Continuous time model

Many species, such as humans and some insects, are best described by a continuous time model of age-structured growth. We will use our experience with the discrete time model to derive the continuous time model. As we will see this model is the basis of many estimates for the *intrinsic rate of increase*, as in Andrewartha and Birch (1954).

We use the following quantities to formulate the model:

- $B(t)$ is the total birth rate in the population.
- $l(x)$ is the fraction of newborn individuals that survive to age x.
- $m(x)$ is the average birth rate for an individual of age x.

To find the total birth rate we sum the births to individuals of all possible ages. This leads to

Recall from your calculus class that integrals are really just sums.

$$B(t) = \int_0^\infty \left(\text{rate of births to parents of age } x\right) dx. \tag{2.54}$$

We now determine the birth rate in equation (2.54), using the following reasoning:

- Rate of births to parents of age x = $m(x)$ times number of individuals of age x.
- Number of individuals of age x = number of births x years ago times fraction of newborn who survive to age x.
- Number of individuals of age x = $B(t - x)l(x)$.

We substitute into equation (2.54) to obtain a single equation relating the birth rate now to the birth rates in the past:

We call this a renewal equation, because it describes how the population is renewed through births.

$$B(t) = \int_0^\infty B(t - x)l(x)m(x)dx. \tag{2.55}$$

We solve this equation under the assumption that the population is in a stable age distribution. Thus, we assume that growth is exponential at the rate r, where we again use the symbol r to represent the intrinsic rate of increase of the population. This can be expressed as

From our analysis of the Leslie matrix formulation, we expect that there will be a stable age distribution.

$$B(t) = e^{rt}B(0). \tag{2.56}$$

Because equation (2.55) also depends on the birth rate at time $t - x$, we rewrite equation (2.56) as:

$$B(t - x) = e^{r(t-x)}B(0). \tag{2.57}$$

Now, substitute these last two relations into the renewal equation (2.55) to obtain

$$e^{rt}B(0) = \int_0^\infty e^{r(t-x)}B(0)l(x)m(x)dx. \tag{2.58}$$

$e^{r(t-x)}/e^{rt} = e^{-rx}$

Now divide both sides by $e^{rt}B(0)$ to arrive at *Euler's equation*:

$$1 = \int_0^\infty e^{-rx} l(x) m(x) dx. \tag{2.59}$$

Note that this equation can be used to find r if we know the life table for a particular organism. However, the birth rates and survival probabilities cannot be measured over continuous time, only at discrete intervals. When this equation is used to estimate r, ecologists usually use a discrete approximation obtained by replacing the integral with a sum

Remember that $\sum_{x=0}^{\infty} e^{-rx} l_x m_x$ means $e^{-0r} l_0 m_0 + e^{-1r} l_1 m_1 + e^{-2r} l_2 m_2 + \ldots$

$$1 = \sum_{x=0}^{\infty} e^{-rx} l_x m_x. \tag{2.60}$$

We have also used the notation l_x and m_x to emphasize that these quantities are defined only at discrete ages. An estimate for r can be found from this equation numerically, as the right-hand side of the equation is monotonic decreasing in r: as r increases the sum always decreases.

We demonstrate this using the data from the vole. By plotting the quantity $\sum_{x=0}^{\infty} e^{-rx} l_x m_x$ against r, we can find the value of r that makes the sum 1. Using this reasoning as shown in Figure 2.5, and then improving the estimate by trying different values of r, we find that the rate of increase for the vole population is approximately 0.7 per 8 weeks.

There is another quantity that is often calculated from the life table, the *net reproduction rate* R_0, which is average total number of (female) offspring produced by a single (female) individual in her lifetime. Try to describe in words why this is given by

$$R_0 = \int_0^\infty l(x) m(x) dx. \tag{2.61}$$

Can you have $r > 0$ and $R_0 < 1$?

What is the relationship between R_0 and r? For the vole population, we find R_0 is 5.90.

Stable age distribution

It is easy to determine the stable age distribution using the framework we have developed here. Denote by $c(x)$ a *density function* for the fraction of individuals of age x in the stable age distribu-

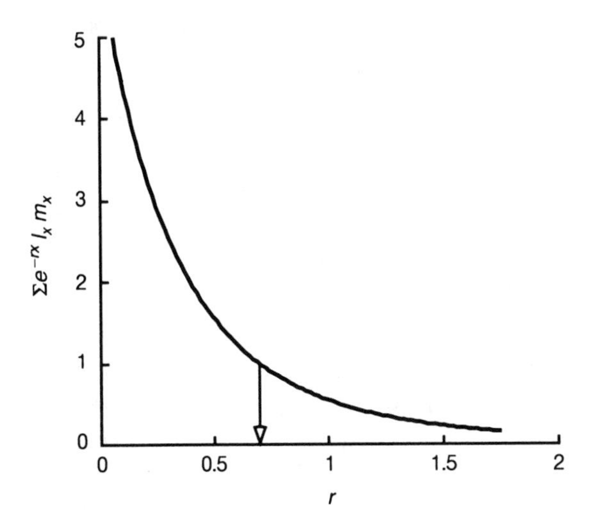

FIGURE 2.5. Plot of $\sum_{x=0}^{8} e^{-rx} l_x m_x$ against different values of r for the vole population whose life table is given in Table 2.2. We find the actual growth rate for this population by finding the value of r that makes the sum 1. As indicated by the arrow on the figure this value is approximately $r = 0.7$.

tion. This means that the fraction of individuals between ages x and $x + dx$ is given by $\int_x^{x+dx} c(z)dz \approx c(x)dx$. Thus, in words

$$c(x) = \frac{\text{number of individuals of age } x}{\text{total number of individuals}}. \tag{2.62}$$

To find a mathematical formula that expresses this descriptive quotient, first note that the total number of individuals is given by

$$\int_0^\infty \big(\text{number of individuals of age } x\big)\, dx. \tag{2.63}$$

One expression for the number of individuals (actually a density function) results from observing that an individual is 10 years old today if it was born 10 years ago and has survived. Thus the number of individuals of age x is given by $B(t - x)l(x)$. Try to determine the reasoning embodied in the following sets of equations:

$$c(x) = \frac{B(t-x)l(x)}{\int_0^\infty B(t-z)l(z)dz} \tag{2.64}$$

At the stable age distribution, $B(t - x) = e^{r(t-x)}B(0)$.

To go from (2.65) to (2.66), divide numerator and denominator by e^{rt}.

TABLE 2.4. Life table and demographic functions for the vole *Microtus agrestis* reared in the laboratory (data from Leslie and Ranson 1940).

age (weeks)	x	l_x	m_x	$l_x m_x$	$e^{-rx} l_x$	c_x	$e^{-rx} l_x m_x$	v_x/v_0
0	0	1.0000	0.0000	0.0000	1.0000	0.5862	0.0000	1.0000
8	1	0.8335	0.6504	0.5421	0.4142	0.2428	0.2694	2.4145
16	2	0.7313	2.3939	1.7507	0.1806	0.1059	0.4323	4.0461
24	3	0.5881	2.9727	1.7482	0.0722	0.0423	0.2145	4.1348
32	4	0.4334	2.4662	1.0689	0.0264	0.0155	0.0652	3.1731
40	5	0.2928	1.7043	0.4990	0.0089	0.0052	0.0151	2.1062
48	6	0.1813	1.0815	0.1960	0.0027	0.0016	0.0030	1.3063
56	7	0.1029	0.6683	0.0687	0.0008	0.0005	0.0005	0.7974
64	8	0.0535	0.4286	0.0229	0.0002	0.0001	0.0001	0.4997
72	9	0.0255	0.3000	0.0076	0.0000	0.0000	0.0000	0.3000

$$= \frac{e^{r(t-x)} B(0) l(x)}{\int_0^\infty e^{r(t-z)} B(0) l(z) dz} \tag{2.65}$$

$$= \frac{e^{-rx} l(x)}{\int_0^\infty e^{-rz} l(z) dz} \tag{2.66}$$

To go from (2.66) to (2.67), approximate the integral by a sum.

$$\approx \frac{e^{-rx} l_x}{\sum_{x=0}^\infty e^{-rx} l_x} \tag{2.67}$$

We can thus find an approximation for the stable age distribution from life table data. The presence of the factor e^{-rx} in (2.67) means that the stable age distribution depends not only on the survivorship rates but also on the intrinsic rate of increase. Equation (2.67) gives the stable age distribution, so by comparing an observed distribution of ages of individuals in a population, one can see if the population is in a stable age distribution.

Why does the intrinsic rate of increase enter into the formula for the stable age distribution? Think of the case of a population that is doubling every year to help answer this question.

Reproductive value

We have seen that a population does not always grow at a rate given by its ultimate growth rate, the growth rate in the stable age distribution. Is there some quantity that always grows at the same rate? The answer to this question arises from a related question posed and answered by Fisher (1930): what is the contribution of an individual of a given age to the future growth of the population? This will depend not only on the expected number of offspring, but also on when they are born. If the population is

This question is related to the concept of present value from economics.

The 'value' of an offspring is defined as the fraction of the total population size it represents – if the population size doubles, the offspring is worth half as much. This concept makes sense from an evolutionary point of view.

growing, then offspring born sooner are more valuable than offspring born later, because an offspring born sooner will represent a larger fraction of the population size. In fact, these offspring should be discounted by a factor that is equal to e^{-rt}, where r is the growth rate of the population and t is how far in the future the offspring are born.

The word 'discount' emphasizes the analogy with economics, where the term discount rate is used. You may be familiar with the idea of paying less than \$50 for a \$50 savings bond that matures in the future.

Reproductive value of an individual at age x is defined relative to the reproductive value of a newborn, as $v(x)/v(0)$. We first see that the probability of survival from age x to age y is $l(y)/l(x)$, so the expected rate of births to an individual of age x when it is y is $m(y)l(y)/l(x)$. We define the reproductive value of an individual of age x as the sum of all future births, where each future birth is weighted by its value relative to a birth now. We actually perform the sum over all ages, where at each age we use the expected number or rate of births. A birth at age y to an individual that is now x will be counted not as 1, but as $e^{-r(y-x)}$. We thus define

$$\frac{v(x)}{v(0)} = \int_x^\infty e^{-r(y-x)} \frac{l(y)}{l(x)} m(y)dy \tag{2.68}$$

$$\approx \sum_{y=x}^{\infty} e^{-r(y-x)} \frac{l_y}{l_x} m_y \tag{2.69}$$

$$= \frac{\sum_{y=x}^{\infty} e^{-ry} l_y m_y}{e^{-rx} l_x}. \tag{2.70}$$

To derive (2.70), notice that we can move the terms involving x outside the summation sign. The resulting equation is easier to use in calculations, as in Table 2.4.

So for the vole example, if $x = 2$, to find the value of v_2/v_0 we use the sum

The notation v_0/v_0 suggests that the value should be 1. Convince yourself that this is true from the formula (2.68).

$$\sum_{y=2}^{9} e^{-ry} l_y m_y = e^{-2r} l_2 m_2 + e^{-3r} l_3 m_3 + \ldots + e^{-9r} l_9 m_9. \tag{2.71}$$

and divide it by $e^{-2r} l_2$.

We have computed the reproductive value for the vole population and listed the values in Table 2.4. From the table we observe some general properties of the reproductive value. The reproductive value typically increases from birth, reaches a maximum between the ages of first and maximal reproduction, and then declines.

Thus, the potential future population size is strongly influenced by the current age distribution of individuals in a population.

Common sense indicates this is true. A human population with an excess (compared to the stable age distribution) of younger individuals will eventually be larger than a population that initially is the same size but starts off with more older individuals.

Problems

1. Evans and Smith (1952) calculated r for the human louse and found it to be approximately 0.1 per day. Use the equation

$$N(t) = N(0)e^{rt} \tag{2.72}$$

and rearrange to get

$$\frac{\ln[N(t)/N(0)]}{r} = t. \tag{2.73}$$

Using this equation, if we know r, $N(0)$, and $N(t)$, we can find t.

Starting with 10 lice, how long will it take for an exponentially growing population of lice to reach 100? 1,000? 100,000,000? 100,000,000,000? Does this surprise you? Discuss.

2. In a population (of an imaginary organism that lives 2 years), the average number of births for 1-year-olds is 1/2, the average number of births for 2-year-olds is 2, and the survival probability from 1 to 2 is 1/2. Death is certain after 2 years.

   (a) Set up a Leslie matrix model for this population.

   (b) If the population starts with 1 adult and no juvenile, find the number of juveniles, adults, and the total population after 3 years. Do the same if the population starts with no adults and 1 juvenile. Who seems to be 'worth' more, in terms of future population size – adults or juveniles?

   (c) Find the long-term growth rate (largest eigenvalue) and the stable age distribution (corresponding eigenvector, the ratio of 1-year-olds to 2-year-olds) for this population. Compare and contrast these numbers to your findings for part (b).

3. Compute

$$\begin{pmatrix} 1 & 2 \\ -3 & 4 \end{pmatrix} \begin{pmatrix} 2 \\ 5 \end{pmatrix}.$$

4. For many bird species, the fecundity and survivorship of adults is independent of the age of the adult bird. Thus, we can think of the population as composed of two classes, juveniles and adults. As in the Leslie matrix, we focus only on females. Set up a model based on a 2 × 2 matrix to describe the dynamics of a bird population without density dependence.

5. Given the following values of l_x and m_x, find r accurate to two decimal places (try different values of r in Euler's equation – if the sum is too large, try a larger value of r; if it is too small, try a smaller value of r):

x	l_x	m_x
0	1.0	0
1	0.6	0
2	0.5	0
3	0.4	3
4	0.3	4
5	0	0

Compute the stable age distribution c_x and the reproductive value v_x for this population, and graph them against the age x. For what age x is v_x maximized? Why?

6. This will require either a spreadsheet or some computer programming. For the snapping turtle life table given in Table 2.3, find r. Then compute the stable age distribution c_x and the reproductive value v_x/v_0 for this population and graph them against the age x. What can you conclude from the graph of the reproductive value?

7. Find an article containing a life table in a recent issue of *Ecology*, *Ecological Applications*, *Journal of Animal Ecology*, *Journal of Ecology*, or *Conservation Biology*. Compute r and compare it to the number in the article. Also, compute the stable age distribution.

Suggestions for further reading

For an in-depth discussion of age-structured population growth, see Caswell (1989). Leslie's (1945, 1948) original papers are worth reading, although they are complex and we now do these computations differently.

An extended discussion of life tables and survivorship curves is in the book by Hutchinson (1978). Numerous examples of life tables are presented in Andrewartha and Birch (1954). Computation of growth rates from life tables has played a role in many discussions in conservation biology; many examples have been discussed in the journal *Conservation Biology*. For example, see the paper by Congdon et al. (1993) on Blanding's turtles.

3

Population Genetics

The famous geneticist Theodosius Dobzhansky once wrote that nothing in biology makes sense except in the light of evolution. This is certainly true about many of the ecological questions of interest to us. Life history patterns, for example, are shaped by selection. The ages at which individuals have offspring, when they disperse, and even when they die are the results of selection. We need to develop some concepts about the genetic properties of populations to help us understand questions in evolutionary ecology. In turn, we will consider the issue of variation within a species – a strictly genetic question – and then turn to questions at the interface between ecology and genetics.

3.1 Genetic questions

What properties of populations can be classified as genetic in nature? Perhaps the most important aspect concerns variation among individuals. One of the primary questions of population genetics is to determine the reasons for the maintenance of *polymorphisms*, two or more different types within a single population. Until the 1960s, the variants studied by geneticists were almost all visible polymorphisms: they could be detected by looking at the organism in question.

One striking example of a visible polymorphism is the existence of two sexes in most higher organisms. There is clearly a mechanism for the maintenance of the polymorphism. If there is only one male in a population and many females, the number of offspring for the male will clearly be higher then the average number of offspring for the females. Thus, any males produced will also have more offspring than any females produced. Consequently, the fraction of males in a population will tend to increase over time. This argument also works in reverse if there is only one female to start with and many males. In general, if a sex becomes rare it is then more 'valuable'.

There are a number of other classic cases of visible polymorphism. One is the maintenance of dark and light forms in the moth *Biston betularia* and the change in the frequency of these forms following an increase in soot from industrial pollution – the case of *industrial melanism*. Another well-known example is the maintenance of the gene responsible for sickle-cell anemia in humans, a disease in which the red blood cells are deformed and less efficient in carrying oxygen. We will explore this case in the problems at the end of the chapter. A third example is the snail *Cepaea nemoralis*, whose numerous shell color and banding polymorphisms have been the subject of extensive study.

What other examples of visible polymorphisms can you think of?

Until the 1960s many geneticists thought genetic variants were relatively rare. Then, beginning with the classic work of Lewontin and Hubby (1966), the technique of *gel electrophoresis* emerged as a method to look for polymorphisms that did not have obvious macroscopic visual effects. Classic gel electrophoresis begins by placing soluble proteins in a gel made of starch. The gel is then subjected to an electric current, and different variants of the same protein may move at different rates, depending on the properties of the amino acids on the outside of the protein. Slices of the gel are treated with the substrates and cofactors catalyzed by a particular enzyme (protein) and a dye that reacts with the end product of the reaction. Proteins migrate under the influence of the electric field and appear as bands on the gel. Bands which travel different distances are different alleles. (Some amino acid

The technique of gel electrophoresis is restricted to looking at soluble proteins that catalyze reactions.

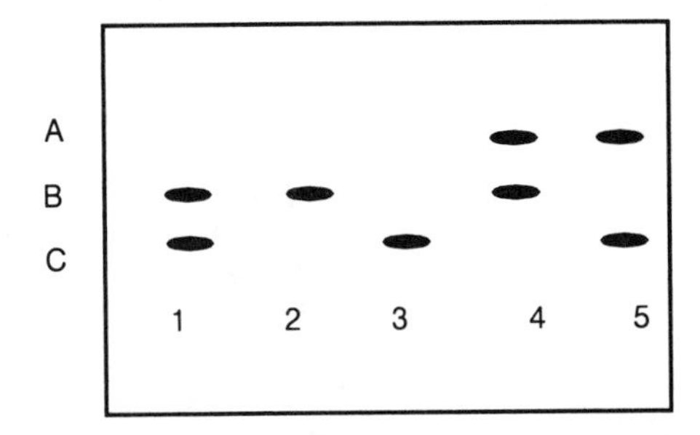

FIGURE 3.1. Schematic diagram illustrating gel electrophoresis as a tool for finding genotypes. Each number corresponds to an individual. After the application of an electric current, a protein moves in the gel, and then the substrates of a reaction catalyzed by that protein as well as a chemical that will react with the end product are placed on the gel. The alleles of the protein, which travel at different rates, show up as bands on the gel. Thus individuals 1 through 5, respectively, have the genotypes BC, BB, CC, AB, and AC. The names of the alleles are chosen arbitrarily.

substitutions which do not result in proteins with different migration rates under electrophoresis will not be detected.) The result of population surveys using gel electrophoresis was the discovery of extensive polymorphisms in almost all species examined.

The question of interest to geneticists became the determination of the forces responsible for the maintenance of the extensive variation revealed by electrophoresis. One possibility is that *selection* (differential reproduction or survival of different genotypes) is the major force responsible for the maintenance of polymorphism. Another possibility is that *drift* (changes in the frequency of different genotypes caused by random sampling effects) coupled with *mutation* (the chance production of new alleles) are the major forces responsible for polymorphism.

During the past decade the techniques used by geneticists to look for polymorphisms have become even more sophisticated. Techniques from molecular biology have been used to show that even the high estimates of levels of polymorphism revealed by gel electrophoresis were lower than those determined by direct or indirect examination of DNA sequences (e.g., Begun and Aquadro, 1995). The major goal remains the determination of the forces responsible for the maintenance of polymorphism.

Genetic techniques may help ecologists to determine rates of exchange of breeding individuals between different populations.

The traits examined through electrophoresis or DNA sequence

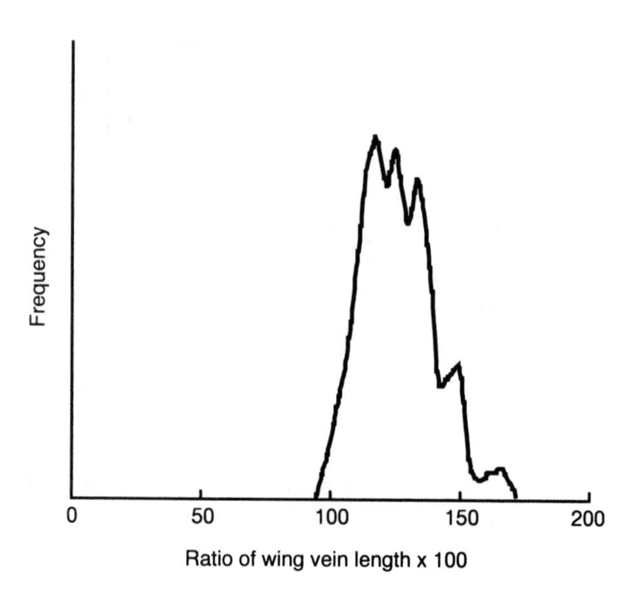

FIGURE 3.2. Distribution of a quantitative trait, the ratio of wing vein lengths in *Drosophila melanogaster* (data from Scharloo et al., 1967). The frequency distribution is very roughly normal.

techniques may not be the ones that are important to an ecologist, in contrast to a trait such as the size of an individual. A very different kind of question emerges when focusing on quantitative characters – the kind that can be measured, such as height, weight, or the ratio of sizes of different body parts (Figure 3.2). Here the issue is why is every individual not the same – presumably optimal – size? Although some variation results from environmental factors, there is also underlying genetic variability in virtually all cases. A related issue is how fast such a quantitative trait evolves in response to selection.

What other quantitative traits can you think of?

3.2 Evolutionary and ecological questions

There are several areas where the interface between ecology and genetics has produced fruitful insights. One involves the use of optimization concepts in ecology. Many ecologists studying the behavior of organisms assume that the behavior is 'optimal'. The natural question is whether it is reasonable to expect that the outcome of evolution is optimal behavior? One specific example is the *evolution of life histories* – when and how reproduction takes place. Another general area of study is *optimal foraging theory* – how a foraging animal searches for patchily distributed food items.

We will look later at comparisons between plants that reproduce many times and those that reproduce once and die.

For ecological questions such as these we may ask:

- Under what circumstances will natural selection produce traits that are optimal?
- Why are all organisms within the population not of the optimal type: why is variability maintained?
- If the environment is changing, or the organism is in a novel environment, one would like to know how fast is the response to selection. How quickly does the population evolve toward the optimum?

Response to changing environments is a question of immense current interest.

Other topics at the interface between ecology and evolution include coevolution, the joint evolution of predator and prey, host and parasitoid, host and pathogen, or plant and pollinator, and the very difficult issue of the process of speciation.

3.3 One-locus model without selection

We will not be able to produce models that will guide our answers to all these questions. The models involved can become very complex, and many of the issues we have raised are still the subject of intense study. Instead, we will study very simple models that may suggest some simple answers. These models will guide our understanding of more complex cases.

Hardy–Weinberg law

At the time of Darwin, the mid-to-late nineteenth century, the factors responsible for the maintenance of variability were not understood. It was unclear why variability was not reduced each generation by the offspring looking like the 'average' of the parents. Gregor Mendel, who deduced the particulate nature of inheritance through his extensive experiments with peas, provided the mechanism – heritable genes – that explained how variability could be maintained. Although Mendel was a contemporary of Darwin, his work was not well known until the early twentieth century. Also, in the early twentieth century, G.H. Hardy, a British mathematician known for his contributions to number theory and analysis, showed mathematically that in the absence of other forces variability is preserved, as we now demonstrate.

We start with the simplest case, a single locus with two alleles. We do not claim that this is realistic.

Consider in a diploid organism a single *locus* with just two *alleles*, A and a. Thus there are three different possible *genotypes* in the population: AA, Aa, and aa. We will make the following assumptions.

Think of how each assumption might be violated. As we formulate the model, consider how the model might have to be changed to accommodate changes in the assumptions.

- Mating is random among individuals in the population.
- There is no selection: the probability of mating and survival is independent of the genotype.
- Generations are nonoverlapping.
- There is no immigration or emigration.
- The population is so large that we can ignore stochastic effects and consider only the frequencies (fractions) of different genotypes.
- There are no mutations at this locus.

What happens to variability in the population at this locus with these assumptions? Let p_{AA} be the frequency of AA individuals, p_{Aa} be the frequency of Aa individuals, and p_{aa} be the frequency of aa individuals. Let the frequency of A alleles be p. If there are N individuals and $2N$ alleles, then the number of A alleles is $2Np_{AA} + Np_{Aa}$. The frequency of A alleles is

There are twice as many alleles as individuals because we are thinking of a diploid organism.

$$p = \frac{2Np_{AA} + Np_{Aa}}{2N} = p_{AA} + \frac{p_{Aa}}{2}. \tag{3.1}$$

The frequency of a alleles is

$$q = 1 - p = \frac{p_{Aa}}{2} + p_{aa}. \tag{3.2}$$

To determine what happens after one generation of mating, we consider all possible matings, their frequency, and the possible offspring and their frequency. These are listed in Table 3.1.

TABLE 3.1. Matings, frequencies, and offspring in a one-locus, two-allele model. The three 'offspring fraction' columns are the fraction of offspring of the mating that are of the given genotype. The 'contribution to next generation' columns are obtained by multiplying the fraction of offspring of each genotype by the frequency of the mating.

	mating	offspring fraction			contribution to next generation		
mating	frequency	AA	Aa	aa	AA	Aa	aa
$AA \times AA$	p_{AA}^2	1	0	0	p_{AA}^2	0	0
$AA \times Aa$	$2p_{AA}p_{Aa}$	1/2	1/2	0	$p_{AA}p_{Aa}$	$p_{AA}p_{Aa}$	0
$AA \times aa$	$2p_{AA}p_{aa}$	0	1	0	0	$2p_{AA}p_{aa}$	0
$Aa \times Aa$	p_{Aa}^2	1/4	1/2	1/4	$p_{Aa}^2/4$	$p_{Aa}^2/2$	$p_{Aa}^2/4$
$Aa \times aa$	$2p_{Aa}p_{aa}$	0	1/2	1/2	0	$p_{Aa}p_{aa}$	$p_{Aa}p_{aa}$
$aa \times aa$	p_{aa}^2	0	0	1	0	0	p_{aa}^2

Summing up the entries in the column 'contribution to next generation', we find the frequency of AA the in next generation:

$$p'_{AA} = p_{AA}^2 + p_{AA}p_{Aa} + p_{Aa}^2/4 \quad (3.3)$$
$$= (p_{AA} + p_{Aa}/2)^2 \quad (3.4)$$
$$= p^2. \quad (3.5)$$

Similar reasoning leads to the conclusion that

$$p'_{aa} = p_{Aa}^2/4 + p_{Aa}p_{aa} + p_{aa}^2 \quad (3.6)$$
$$= (p_{Aa} + p_{aa}/2)^2 \quad (3.7)$$
$$= q^2 \quad (3.8)$$

and

$$p'_{Aa} = 1 - p^2 - q^2 \quad (3.9)$$
$$= 2pq. \quad (3.10)$$

Equation (3.10) follows from the observation that $1 = p + q$ implies that $1 = (p+q)^2 = p^2 + 2pq + q^2$.

Thus, after one generation, no matter what the initial genotypic frequencies were, the genotypic frequencies are given by

Note that the frequencies in (3.11) through (3.13) sum to 1.

$$p_{AA} = p^2 \quad (3.11)$$
$$p_{Aa} = 2pq \quad (3.12)$$
$$p_{aa} = q^2. \quad (3.13)$$

We draw several very important conclusions from this result. First, after one generation the genotype frequencies are completely determined by the initial allele frequencies. This also means that we can express our models in terms of the single quantity, p, the frequency of A, which is a great simplification. Second, the allele frequencies and genotype frequencies remain constant from generation to generation after the first generation; genetic variability is not eliminated. Third, any deviation from the genotype frequencies (3.13), which are called the *Hardy-Weinberg frequencies* to honor the mathematicians who derived the equations, must result from violation of one of our assumptions. Unfortunately, in practice it is very difficult to detect statistically deviations from Hardy-Weinberg frequencies, because none of the available statistical tests are very sensitive.

3.4 One-locus model with selection

At this point we can begin exploring the consequences of changing any of the assumptions we have made in our initial model. Our initial model predicts that variability would be maintained, but does not provide any reason why a particular allele frequency should be found. Also, some variability will clearly be lost because of random sampling, so we need to look at forces that can truly maintain variability. We thus turn to an examination of the role of selection. In this context, we are also motivated by our interest in understanding the dynamics of traits of ecological interest.

We thus add to the simple one-locus model the complicating factor of selection and ask:

Measuring selection directly can be very difficult because very small differences at the individual level in survivorship probability or reproduction can have large effects at the population level.

- How strong must selection be to produce an observed change in allele frequencies? We will use the answer to this question to consider the case of 'industrial melanism', the rapid increase in the frequency of dark forms of a forest moth following the presence of soot on trees.

- Under what conditions will a polymorphism be maintained? We will examine the case of sickle-cell anemia in the problems.

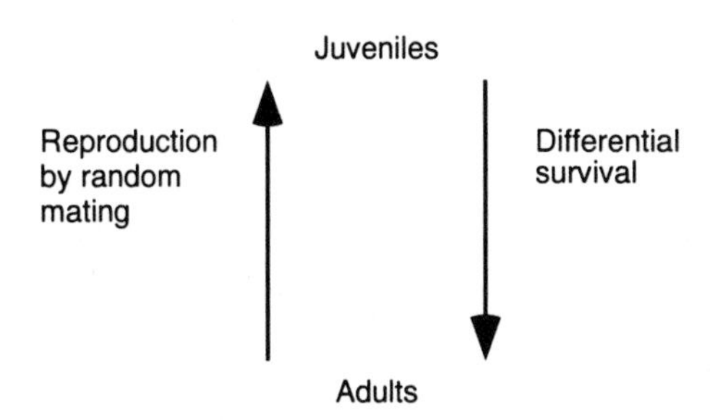

FIGURE 3.3. Life cycle used in the model

Fitness

To answer these questions, we must talk about *fitness* and *selection*. The fitness of an individual is its probability of leaving behind offspring as determined through differential reproduction, survival, and possibly representation among gametes, depending on the phenotype. We will temporarily equate fitness with survival probability, or viability. By selection we mean the process whereby the more fit individuals are chosen.

Viability selection

In addition to the assumptions made under Hardy–Weinberg circumstances, we will make further assumptions to obtain a simple model incorporating selection. We assume that viabilities do not change through time. We also ignore any frequency dependence in fitnesses; fitnesses of each genotype are assumed to be independent of the current makeup of the population. The two stages in the model are illustrated in Figure 3.3.

Can you think of cases in which frequency-dependent selection is likely?

The model is described in Table 3.2. At the initial juvenile stage we can assume that the frequencies are given by the Hardy–Weinberg proportions, because the juveniles resulted from a cycle of mating and we have already shown that only one generation is required to establish Hardy–Weinberg proportions. To find the *relative* adult proportions, we multiply the juvenile frequencies by the relative survival rates. To convert these relative proportions to frequencies, we divide by their sum, the (relative) mean

You can add the entries in the last row of Table 3.2 to see that they sum to 1.

TABLE 3.2. The one-locus, two-allele model with selection.

	genotype		
	AA	Aa	aa
juvenile frequencies	p^2	$2pq$	q^2
relative survival rates	w_{AA}	w_{Aa}	w_{aa}
relative adult frequencies	$p^2 w_{AA}$	$2pqw_{Aa}$	$q^2 w_{aa}$
adult frequencies	$p^2 w_{AA}/\overline{w}$	$2pqw_{Aa}/\overline{w}$	$q^2 w_{aa}/\overline{w}$

fitness:

$$\overline{w} = p^2 w_{AA} + 2pqw_{Aa} + q^2 w_{aa}. \tag{3.14}$$

We then use these results to determine the frequency of A alleles in adults, which is the same as the frequency of A alleles in the juveniles of the next generation, p'. From the last row of Table 3.2 we see that

$$\begin{aligned} p' &= p'_{AA} + p'_{Aa}/2 && (3.15)\\ &= p^2 w_{AA}/\overline{w} + (1/2)2pqw_{Aa}/\overline{w} && (3.16)\\ &= \frac{p(pw_{AA} + qw_{Aa})}{\overline{w}} && (3.17)\\ &= \frac{pw_A}{\overline{w}}, && (3.18) \end{aligned}$$

Both by guessing, and by following similar steps, find the equation for q'.

where in the last line we have defined the mean fitness of allele A to be

$$w_A = pw_{AA} + qw_{Aa}. \tag{3.19}$$

Does this make sense? If we know that an individual has a single A allele, then with probability p its other allele is A, while with probability q its other allele is a. Thus the mean relative fitness of an individual carrying an A allele is $pw_{AA} + qw_{Aa}$. From equation (3.18) we see that allele A increases in frequency if the mean fitness of individuals carrying allele A is greater than the mean fitness of the population.

Before using this model to examine some of the questions posed earlier, we phrase the model in terms of a different set of parameters that simplify the algebra and make our conclusions more apparent. Let

$$w_{AA} = 1 - s \quad (3.20)$$

$$w_{Aa} = 1 \quad (3.21)$$

$$w_{aa} = 1 - t. \quad (3.22)$$

At this point, the sign of s and t is arbitrary. With this choice of parameters, we see that

Can you give a heuristic discussion of the meaning of s and t?

Remember to use the fact that $p + q = 1$.

$$w_A = p(1 - s) + q = p + q - ps = 1 - ps \quad (3.23)$$

$$w_a = p + q(1 - t) = p + q - qt = 1 - qt \quad (3.24)$$

and that

Remember that $1 = (p + q)^2 = p^2 + 2pq + q^2$.

$$\overline{w} = p^2(1 - s) + 2pq + q^2(1 - t) \quad (3.25)$$

$$= p^2 + 2pq + q^2 - p^2s - q^2t \quad (3.26)$$

$$= 1 - p^2s - q^2t \quad (3.27)$$

$$= p(1 - ps) + q(1 - qt). \quad (3.28)$$

Equilibria

We are now ready to begin to answer the first of the questions posed earlier: when is a polymorphism maintained at *equilibrium*? The notion of an equilibrium is central to much of the rest of the text. An equilibrium is a value for a variable that can remain constant through time.

Let Δp be the change in p, the frequency of A, from generation to generation. Then

Compare the development here to the graphical approach used in Figures 3.4 through 3.7 below.

$$\Delta p = p' - p = \frac{pw_A}{\overline{w}} - p. \quad (3.29)$$

Denote the equilibrium value of p by $\hat{p}$. At an equilibrium, the change in allele frequencies from generation to generation is zero, so set $\Delta p = 0$:

$$0 = \frac{\hat{p}w_A}{\overline{w}} - \hat{p} \quad (3.30)$$

$$= \frac{\hat{p}(1 - \hat{p}s)}{1 - \hat{p}^2s - \hat{q}^2t} - \hat{p} \quad (3.31)$$

$$= \frac{\hat{p}(1 - \hat{p}s) - \hat{p}(1 - \hat{p}^2s - q^2t)}{1 - \hat{p}^2s - \hat{q}^2t}. \quad (3.32)$$

Box 3.1. Finding the equilibrium of a discrete time model with a single variable.

To find the equilibrium of the discrete time model

$$p' = f(p)$$

write the change in p, δp, as

$$\delta p = f(p) - p.$$

Set δp to be zero, and find the equilibrium, $\hat{p}$, as the solution of the resulting equation:

$$0 = f(\hat{p}) - \hat{p}.$$

We continue by setting the numerator of (3.32) to zero:

$$0 = \hat{p}(1 - \hat{p}s) - \hat{p}(1 - \hat{p}^2 s - \hat{q}^2 t) \quad (3.33)$$

$$= \hat{p}[1 - \hat{p}s - (1 - \hat{p}^2 s - \hat{q}^2 t)]. \quad (3.34)$$

Thus either

$$\hat{p} = 0 \quad (3.35)$$

We get this pair because if the right-hand side of (3.34) is zero, one or the other of the factors must be.

or

$$1 - \hat{p}s - (1 - \hat{p}^2 s - \hat{q}^2 t) = 0. \quad (3.36)$$

It makes sense that $\hat{p} = 0$ is an equilibrium, because if there are no A alleles in one generation and there is no mutation or immigration (as we have assumed), there will be no A alleles the next generation. Similar reasoning suggests that $\hat{p} = 1$ must be an equilibrium as well and thus $1 - \hat{p}$ must be a factor of (3.36). We therefore notice that (using $\hat{q} = 1 - \hat{p}$)

We are trying to divide by the factor $1 - p$.

$$0 = 1 - \hat{p}s - (1 - \hat{p}^2 s - \hat{q}^2 t) \quad (3.37)$$

$$= 1 - \hat{p}s - 1 + \hat{p}^2 s + (1 - \hat{p})^2 t \quad (3.38)$$

$$= -\hat{p}s + \hat{p}^2 s + (1 - \hat{p})^2 t \quad (3.39)$$

$$= (1 - \hat{p})[-\hat{p}s + (1 - \hat{p})t] = 0. \quad (3.40)$$

We conclude that $\hat{p} = 0$, or $\hat{p} = 1$ or that

$$-\hat{p}s + (1 - \hat{p})t = 0. \quad (3.41)$$

Solving for $\hat{p}$, we find:

$$-\hat{p}s - \hat{p}t + t = 0 \tag{3.42}$$

$$t = \hat{p}s + \hat{p}t. \tag{3.43}$$

We finally conclude that

$$\hat{p} = \frac{t}{s+t}. \tag{3.44}$$

Note that for (3.44) to represent a polymorphic equilibrium (an equilibrium with both alleles present) we must have $0 < \hat{p} < 1$, which clearly requires that either s and t are both negative or both positive. *Thus a polymorphic equilibrium is possible only if the heterozygote is the most fit or the least fit genotype.*

Thus this model can have either two or three possible equilibria.

What does (3.44) imply about $\hat{p}$ if $s > 0$ and $t < 0$?

Stability of equilibria

We have discovered that the one-locus model can have up to three equilibria. We now know what the outcome is if the allele frequencies are exactly zero, or exactly one, or exactly $t/(s+t)$. But, what happens if the population starts at an allele frequency that is not one of the equilibria? We will use an argument, based on approximations near an equilibrium, that we repeat several times in this text.

We begin by going through a series of steps that in fact justify the simpler procedure we will use in general.

We first start near the $\hat{p} = 0$ equilibrium. We first look at what happens if we start with no A alleles and then introduce a small number of A alleles, as would happen if there was a low mutation rate or a small number of immigrants with A. We thus are looking at the case where p is very small. We begin by writing out the equation for p' in terms of p only, eliminating q.

The equilibrium $\hat{p} = 0$ corresponds to a population with no A alleles.

$$p' = \frac{p(1-ps)}{1-p^2s-q^2t} \tag{3.45}$$

$$= \frac{p-p^2s}{1-p^2s-(1-p)^2t} \tag{3.46}$$

$$= \frac{p-p^2s}{1-p^2s-t+2pt-p^2t}. \tag{3.47}$$

We are ready to use our assumption that p is very small. Notice that if $p = 10^{-2}$, then $p^2 = 10^{-4}$, so if p is very small we can approximate the numerator by p because the other term will be much smaller. Similarly, in the denominator the term $1-t$ will be much larger than any of the other terms, all of which contain p. Thus, we approximate (3.47) as

$$p' \approx \frac{p}{1-t}. \tag{3.48}$$

We are ignoring the special case in which t is precisely 1.

Do you recognize this equation? It is identical to the equation for discrete time exponential growth which we studied earlier. Thus, we identify two distinct cases:

- If $t > 0$, then $\frac{1}{1-t} > 1$ so p increases when it is small.
- If $t < 0$ then $\frac{1}{1-t} < 1$, so p decreases when it is small.

Does this conclusion make sense? If p is small and A is rare, there will be so few AA individuals that we can ignore them. If the heterozygote Aa is more fit than the homozygote aa, we would expect that A would increase in frequency. A more fit heterozygote corresponds to $t > 0$.

We can use a similar argument to show that if q is small, or p is near 1, then:

- If $s > 0$, then q increases (p decreases).
- If $s < 0$, then q decreases (p increases).

We have just determined when each of the equilibria $\hat{p} = 0$ and $\hat{p} = 1$ is stable. An equilibrium is *stable* if a population starting near the equilibrium approaches the equilibrium. The notion of stability is very important, because we expect to find populations near stable equilibria and do not expect to find populations near unstable equilibria because only stable equilibria are approached as time increases.

Give a definition of an unstable equilibrium.

A simpler way of computing stability

We are still justifying, and not simply describing, the procedure we will use in practice to find stability.

We would not want to go through all these steps every time we want to compute stability. Luckily, there is a different way of viewing the calculation we have just done. Think of p' as a function of p, $p'(p)$.

If we wish to approximate p' near the equilibrium $p = 1$, we can use a Taylor series (Box 3.2). The 'variable' in the Taylor series

Box 3.2. Taylor series of a function of a single variable.

We often need to approximate a function in our analysis of models. The function $f(x)$ can be approximated near the value x_0 using a *Taylor series* as $f(x_0)$ plus the difference between x and x_0 multiplied by how much the function f changes as its argument, x, changes:

$$f(x) \approx f(x_0) + (x - x_0) \left. \frac{df}{dx} \right|_{x=x_0} .$$

When we use a Taylor series in this book, x_0 will be an equilibrium, so we will have a value for $f(x_0)$.

will be $p - \hat{p} = p - 1$. The Taylor series is

$$p' \approx p'\big|_{p=1} + (p - 1) \left. \frac{dp'}{dp} \right|_{p=1} . \tag{3.49}$$

We know that $p = 1$ is an equilibrium, so p' evaluated at $p = 1$ is just this equilibrium value, 1. Rewrite this equation (replacing the approximately equal by an equal sign) as

$$p' - 1 = (p - 1) \left. \frac{dp'}{dp} \right|_{p=1} . \tag{3.50}$$

We compute the derivative of p' from equation (3.45):

$$\frac{dp'}{dp} = \frac{d\left(\frac{p(1-ps)}{1-p^2s-q^2t}\right)}{dp} \tag{3.51}$$

$$= \frac{(1 - 2ps)(1 - p^2s - q^2t) - (p - p^2s)(-2ps + 2qt)}{(1 - p^2s - q^2t)^2} \tag{3.52}$$

The formula for the derivative of u/v is $\left(\frac{du}{dp}v - u\frac{dv}{dp}\right)/v^2$.

We substitute $p = 1$ in this expression to find:

We make this substitution because we are interested in the dynamics near $p = 1$.

$$\left. \frac{dp'}{dp} \right|_{p=1} = \frac{(1 - 2s)(1 - s) - (1 - s)(-2s)}{(1 - s)^2} \tag{3.53}$$

$$= \frac{1}{1 - s} . \tag{3.54}$$

We substitute (3.54) into (3.50) to get:

$$p' - 1 = (p - 1)\left(\frac{1}{1 - s}\right). \tag{3.55}$$

We now give a name to the deviation from the equilibrium, $p - 1$, by setting $\delta p = p - 1$. We notice that the deviation in the next generation, $\delta p'$, is $p' - 1$. So our equation takes the nice form

$$\delta p' = \delta p\left(\frac{1}{1 - s}\right). \tag{3.56}$$

From this equation we can reach the same conclusion as before about stability.

We can now write down the general scheme for what we have done. To approximate p' at the equilibrium $p = \hat{p}$, we can use a Taylor series:

$$p' \approx p'\big|_{p=\hat{p}} + (p - \hat{p})\left.\frac{dp'}{dp}\right|_{p=\hat{p}}. \tag{3.57}$$

We know that $\hat{p}$ is an equilibrium, so p' evaluated at $p = \hat{p}$ is just this equilibrium value, $\hat{p}$. Rewrite this equation (replacing the approximately equal by an equal sign) as

$$p' - \hat{p} = (p - \hat{p})\left.\frac{dp'}{dp}\right|_{p=\hat{p}}. \tag{3.58}$$

We now give a name to the deviation from the equilibrium, $p - \hat{p}$, by setting $\delta p = p - \hat{p}$. Notice that the deviation in the next generation, $\delta p'$, is $p' - \hat{p}$. So our equation for the dynamics of the deviation from equilibrium takes the nice form

$$\delta p' = \delta p\left.\frac{dp'}{dp}\right|_{p=\hat{p}}. \tag{3.59}$$

From this equation we can easily determine the stability of the equilibrium, reaching the same conclusion as before.

- If $\left.\frac{dp'}{dp}\right|_{p=\hat{p}} > 1$, or $\left.\frac{dp'}{dp}\right|_{p=\hat{p}} < -1$, then the equilibrium is unstable.
- If $-1 < \left.\frac{dp'}{dp}\right|_{p=\hat{p}} < 1$, then the equilibrium is stable.

The procedure is summarized in Box 3.3.

Box 3.3. Determining the stability of an equilibrium of a discrete time model with a single variable.

To find the stability of the equilibrium $\hat{p}$ of the discrete time model

$$p' = f(p)$$

compute the derivative

$$\left.\frac{df}{dp}\right|_{p=\hat{p}} \tag{a}$$

and evaluate it at the equilibrium. If this quantity is greater than 1 (in absolute value), the equilibrium is unstable; if this quantity is less than 1 (in absolute value), the equilibrium is stable. (Although the quantity in (a) is positive for all the models in this chapter, for some ecological models it can be negative, and we have to allow for this possibility. Reasoning similar to that used in the discrete time, density-independent growth model would show that instability would result if the 'growth rate' (a) were less than -1.)

Graphical approach

Before summarizing the biological interpretation of our results, we present one more way of understanding the dynamics of the one-locus, one-allele model. We will look at graphs of the change in allele frequency from equation (3.29), $\Delta p = p' - p$, against p. On these graphs, equilibria are values where $\Delta p = 0$, and p increases if Δp is positive and decreases if Δp is negative. So, by using the graphs to find changes in p, we can not only find equilibria but also determine their stability, since it is easy to determine the sign of the change of the allele frequency.

As we see in the next chapter, we need to be cautious in applying this approach in general. If Δp is too large, the variable can 'overshoot' the equilibrium and end up farther away and never reach the equilibrium. This complication does not occur in one-locus population genetics models, but is found in the ecological models we look at later.

We apply this graphical approach to four different cases. The behavior in these four cases can be read off Figures 3.4 through 3.7.

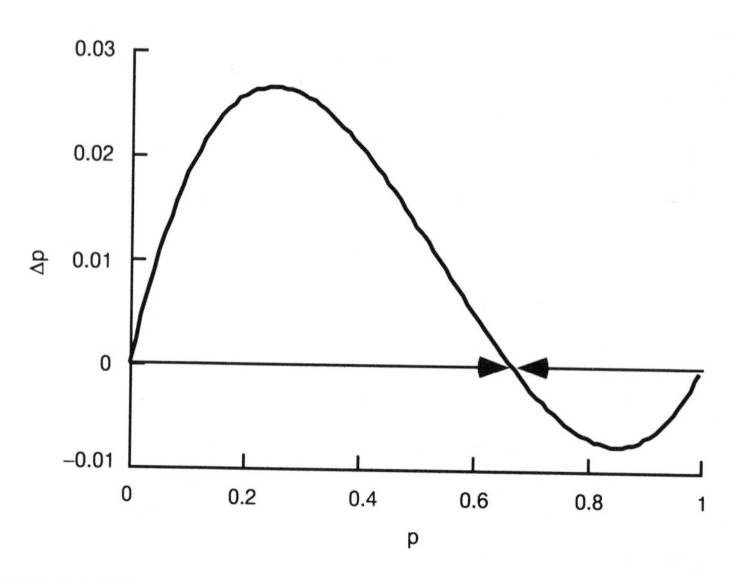

FIGURE 3.4. Allele frequency changes in the one-locus model, with the heterozygote most fit. The change in allele frequency is plotted against the allele frequency. When $\Delta p > 0$, p increases; when $\Delta p < 0$, p decreases. The arrows along the line $\Delta p = 0$ indicate the direction of the change in allele frequency. The fitnesses are $w_{AA} = 0.9$; $w_{Aa} = 1.0$; $w_{aa} = 0.8$. The stable polymorphism with $p = 2/3$ is approached from all initial conditions except fixation.

Conclusions about equilibria

From the analysis we have just performed, we can identify four cases corresponding to different selection schemes and resulting equilibrium behavior of alleles, as listed in Table 3.3. Some typical dynamics of allele frequencies in these cases are graphed in Figures 3.8 through 3.10. The reason for doing the analysis rather than just looking at the computer solutions in the figures is that the analysis gives results for all possible fitness combinations (those that have a heterozygote as most fit, for example), while Figure 3.8 only gives results for a particular set of fitnesses. We would have to use the computer on an infinite number of possible combinations to reach the same conclusions as we can through the analysis.

The behavior of the four cases is summarized as follows.

CASE A. The heterozygote is most fit, and there is a stable polymorphic equilibrium.

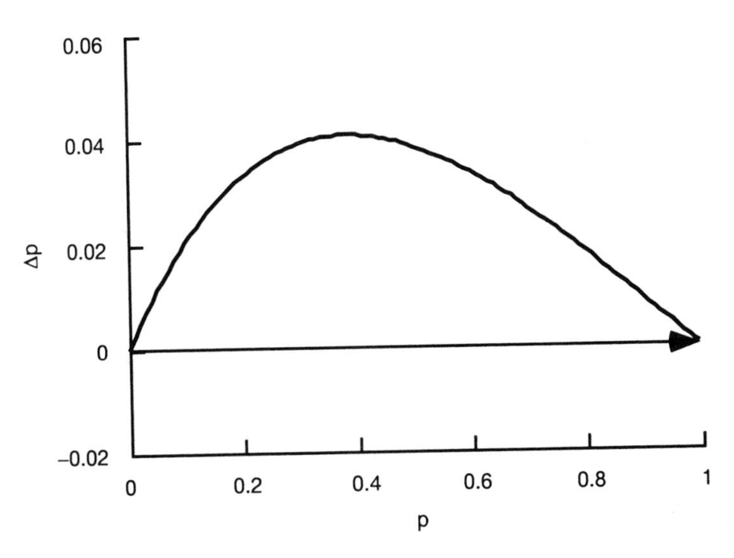

FIGURE 3.5. Allele frequency changes in the one-locus model, with allele A favored. The change in allele frequency is plotted against the allele frequency. When $\Delta p > 0, p$ increases; when $\Delta p < 0, p$ decreases. The arrow along the line $\Delta p = 0$ indicates the direction of the change in allele frequency. The fitnesses are $w_{AA} = 1.1; w_{Aa} = 1.0; w_{aa} = 0.8$. The equilibrium with A fixed is approached.

CASE B. The allele A is favored and always increases in frequency.

CASE C. The allele a is favored and always increases in frequency.

CASE D. The heterozygote is least fit, and the polymorphic equilibrium is unstable. The outcome – which allele eventually becomes fixed (reaches frequency 1) – depends on the starting conditions.

An important observation is that when the heterozygote is most fit, the evolution is constrained by the genetic system in the sense that a population consisting entirely of heterozygotes would have the highest fitness. However, this is never achieved because a population composed entirely of heterozygotes would produce some homozygotes the following generation.

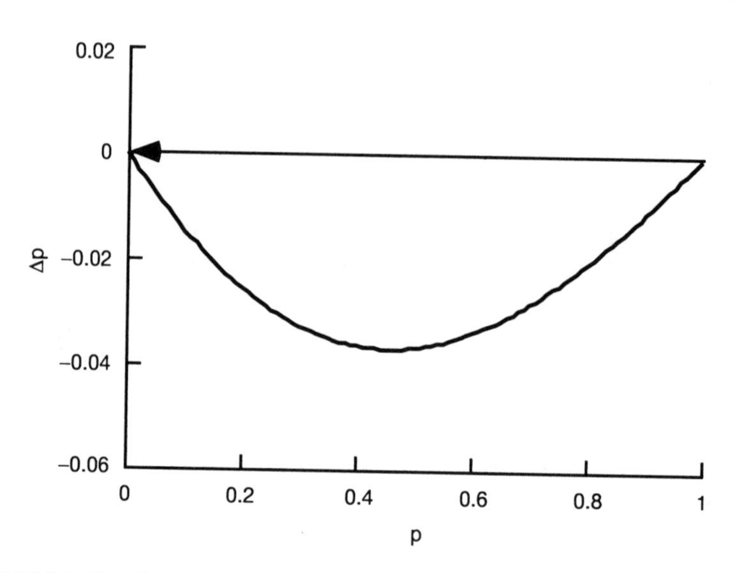

FIGURE 3.6. Allele frequency changes in the one-locus model, with allele a favored. The change in allele frequency is plotted against the allele frequency. When $\Delta p > 0, p$ increases; when $\Delta p < 0, p$ decreases. The arrow along the line $\Delta p = 0$ indicates the direction of the change in allele frequency. The fitnesses are $w_{AA} = 0.9; w_{Aa} = 1.0; w_{aa} = 1.2$. The equilibrium with a fixed is approached.

TABLE 3.3. Equilibrium behavior of the one-locus, two-allele model with selection as a function of the fitness of the homozygotes.

	$s > 0$	$s < 0$
$t > 0$	CASE A. p increases when small, q increases when small, polymorphic equilibrium exists.	CASE B. p increases when small, q decreases when small, no polymorphic equilibrium exists.
$t < 0$	CASE C. p decreases when small, q increases when small, no polymorphic equilibrium exists.	CASE D. p decreases when small, q decreases when small, polymorphic equilibrium exists.

Allele frequency dynamics with one locus

Industrial melanism

The case of industrial melanism is well known. More than 70 species of light-colored moths in Britain have dark variants which increased from very low frequencies after industrialization increased the amount of soot on trees. The dark forms were more cryptic

Dominant alleles produce the same phenotype whether one or two copies are present. Thus a dominant allele causing melanism will lead to a dark moth in the heterozygote, whereas with a recessive pigment allele only the homozygote will be dark.

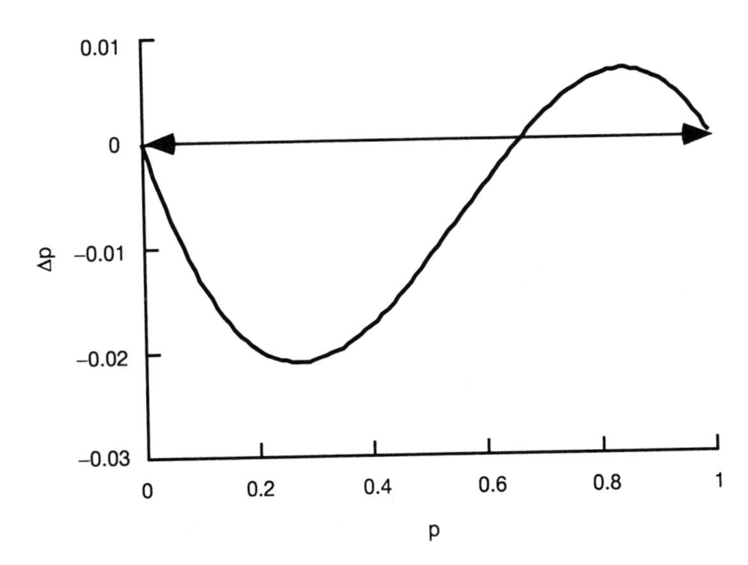

FIGURE 3.7. Allele frequency changes in the one-locus model, with the heterozygote least fit. The change in allele frequency is plotted against the allele frequency. When $\Delta p > 0$, p increases; when $\Delta p < 0$, p decreases. The arrows along the line $\Delta p = 0$ indicate the direction of the change in allele frequency. The fitnesses are $w_{AA} = 1.1$; $w_{Aa} = 1.0$; $w_{aa} = 1.2$. The two fixation equilibria are approached, with the outcome depending on the initial allele frequency.

and less subject to predation by birds. What is striking is that in virtually all cases the alleles responsible were dominant, although recessive alleles causing melanism exist. Why? To answer this question we will look at allele frequency changes through time.

First, look at the dynamics of q, the frequency of a. From our earlier work , we see that

$$\Delta q = q' - q = \frac{q - q^2 t}{1 - q^2 t - p^2 s} - q \tag{3.60}$$

$$= \frac{-q^2 t + q^3 t + p^2 q s}{1 - q^2 t - p^2 s}. \tag{3.61}$$

We first assume that A is dominant and a is recessive so that $s = 0$ and $t \neq 0$. This assumption implies that

$$\Delta q = \frac{-q^2 t - q^3 t}{1 - q^2 t}. \tag{3.62}$$

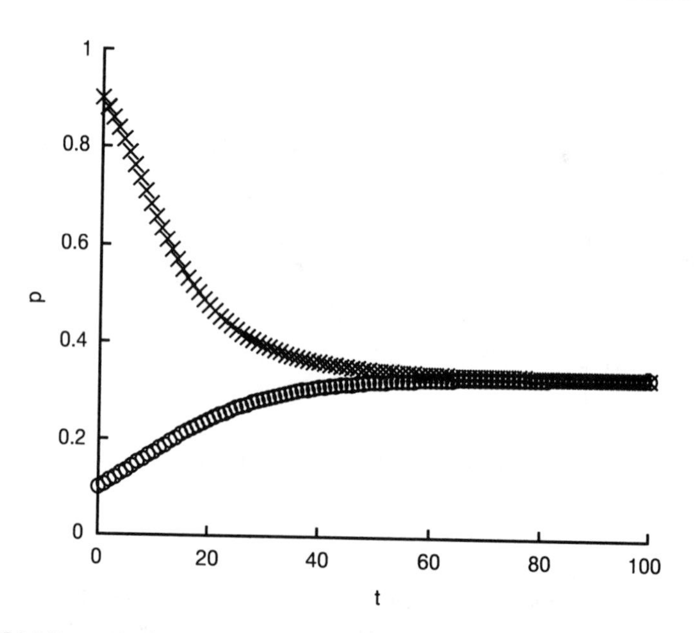

FIGURE 3.8. Dynamics of the one-locus model, with the heterozygote most fit. The frequency of A is plotted against time for two different initial conditions. The fitnesses are $w_{AA} = 0.8$; $w_{Aa} = 1.0$; $w_{aa} = 0.9$. The stable polymorphism with $p = 1/3$ is approached from the two different initial conditions.

We ignore the q^3t in the numerator because it is small compared to q^2t, and the q^2t in the denominator because it is small relative to 1.

Notice that when the a allele is rare in the population, q is small, and (3.62) reduces to

$$\Delta q = -q^2 t. \tag{3.63}$$

This contrasts with the case where A is recessive and a is dominant so that $s \neq 0$ and $t = 0$. Then

$$\Delta q = \frac{p^2 qs}{1 - p^2 s}. \tag{3.64}$$

We replace $p = 1 - q$ by 1 because q is small.

Here, if q is small (3.64) reduces to

$$\Delta q = \frac{qs}{1 - s}. \tag{3.65}$$

Notice that Δq is much larger for small values of q in the case where a is dominant; the right-hand side of (3.65) is much bigger than the right-hand side of (3.63) because a term with just q is much larger than one with q^2 when q is small. The biological explanation is that if a is rare, it is present almost exclusively in

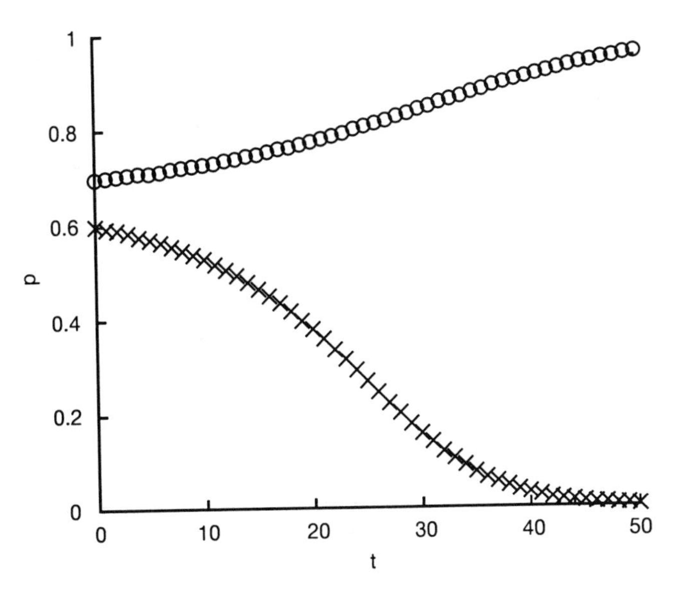

FIGURE 3.9. Dynamics of the one-locus model with the heterozygote least fit. The frequency of A is plotted against time for two different initial conditions. The fitnesses $w_{AA} = 1.0$; $w_{Aa} = 0.9$; $w_{aa} = 1.1$ are used. Depending on the initial conditions, the population approaches fixation for A or a.

heterozygotes, so selection proceeds much more rapidly when the effects of a are observed in heterozygotes. This explains why virtually all the cases of industrial melanism resulted from dominant alleles. The rate of increase for a recessive allele would be too slow when it was rare to explain the relatively rapid appearance of melanic forms in response to the change in the level of soot on trees. The effect is dramatic, as illustrated for specific fitness values in Figure 3.11.

Melanic alleles were always present in very low frequency before the advent of industrialization.

Recessive lethals

In typical populations of *Drosophila melanogaster*, one third of the chromosomes have one or more recessive lethals (Salceda, 1977). Why do populations contain a significant number of alleles that are recessive lethals – why have these not been eliminated by selection? We can answer this question by considering the case of strong selection, focusing on recessive lethals. Assume that aa

Here we are looking at recessive lethals that are lethal when homozygous, but where the heterozygote has essentially the same fitness as the more fit homozygote.

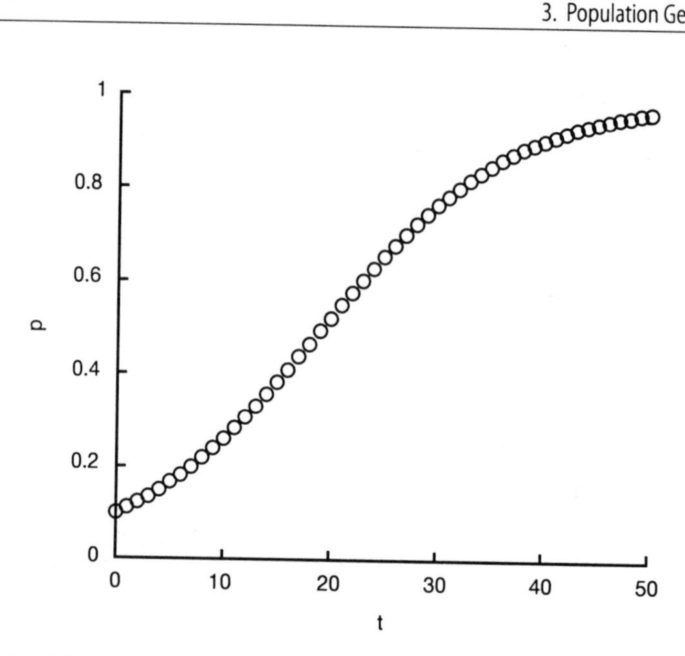

FIGURE 3.10. Dynamics of the one-locus model with the allele A favored. The frequency of A is plotted against time. The fitnesses are $w_{AA} = 1.0$; $w_{Aa} = 0.9$; $w_{aa} = 0.8$. The population approaches fixation of A.

is lethal, so $t = 1$. Then (3.62) becomes:

$$\Delta q = \frac{-q^2(1-q)}{1-q^2} \tag{3.66}$$

$$= \frac{-q^2(1-q)}{(1-q)(1+q)} \tag{3.67}$$

$$= \frac{-q^2}{(1+q)}. \tag{3.68}$$

As illustrated in Table 3.4 and Figure 3.12, the elimination of a recessive lethal proceeds very slowly, once it becomes rare. Thus, it is not surprising that there are many recessive lethals in natural populations. An experiment illustrating the slow elimination of recessive lethals is shown in Figure 3.13.

Mutation

We have seen that selection favoring a heterozygote can be one force maintaining variability in a population. Here we will explore another process that maintains variability – the balance between

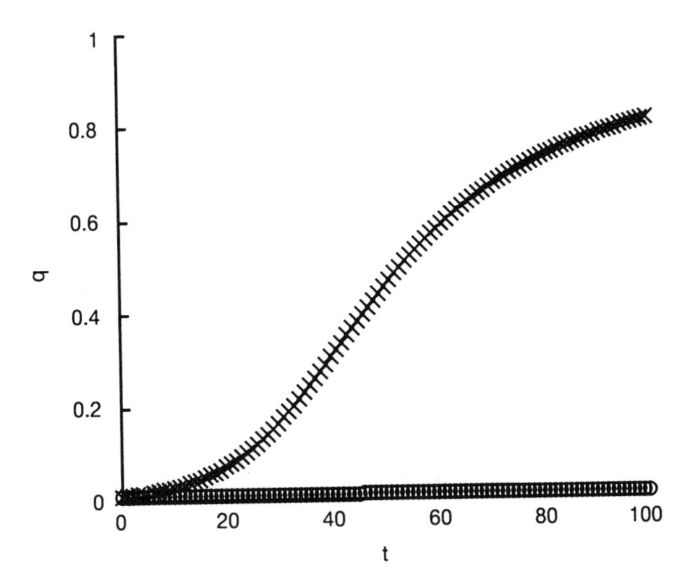

FIGURE 3.11. Dynamics of the one-locus model with the a allele favored : comparison between recessive and dominant cases. The frequency of a, q, is plotted against time for the two cases. When the a allele is dominant, with fitnesses $w_{aa} = 1.0$; $w_{Aa} = 1.0$; $w_{AA} = 0.9$, the rate of increase is very rapid. In contrast, if the a allele is recessive, with fitnesses $w_{aa} = 1.0$; $w_{Aa} = 0.9$; $w_{AA} = 0.9$, the rate of increase of a is imperceptibly slow.

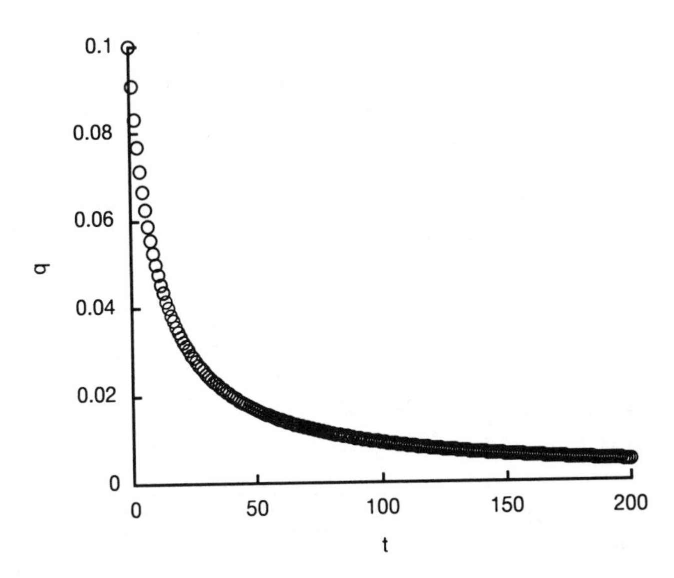

FIGURE 3.12. Model dynamics of selection against a recessive lethal. Note that when the lethal becomes rare, selection proceeds very slowly.

TABLE 3.4. Change in frequency of a recessive lethal in one generation.

q	Δq
0.1	−0.0091
0.01	−0.000099
0.001	−0.000001

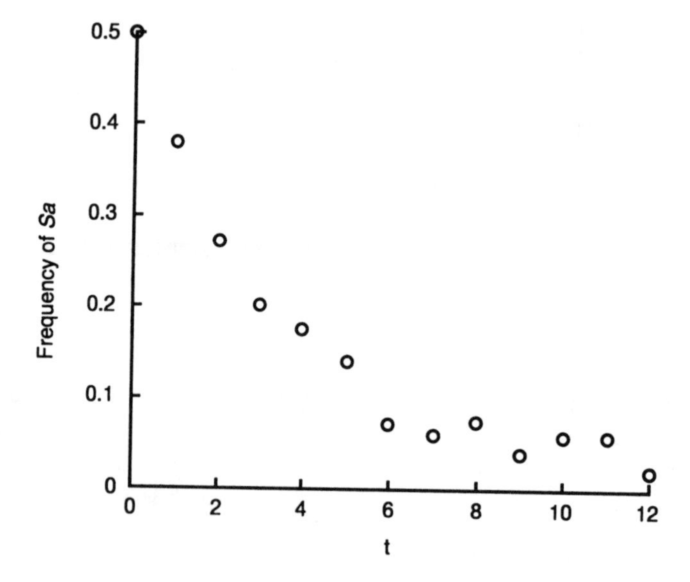

FIGURE 3.13. Dynamics of selection against a recessive lethal, *Sa*, in one line of an experiment on *Tribolium castaneum* (data from Dawson, 1970). The population size is small, so the action of drift is one explanation for differences between model (Figure 3.12) and experiment.

TABLE 3.5. Parameters used in the mutation–selection balance model. The degree of dominance of a is measured by h. When $h = 1$, a is completely dominant, while if $h = 0$, a is completely recessive.

w_{AA}	w_{Aa}	w_{aa}
1	$1 - hs$	$1 - s$

mutation and selection. We will phrase our model using slightly different parameters, as given in Table 3.5. We introduce the new parameter h, which measures the degree of dominance of the allele a. When $h = 1$, a is dominant; when $h = 0$, a is recessive.

We will seek to determine what levels of mutation and selection are necessary for maintaining an observed level of polymorphism. We assume that A mutates to a at the rate μ per generation per allele. We ignore the role of mutation from a to A because we will look at cases where a is relatively rare.

To determine the outcome of a balance between mutation and selection, we set the sum of the change in allele frequency caused by mutation plus the change in allele frequency from selection equal to zero and solve for the allele frequency. Typical mutation rates are small, of the order of 10^{-5} per allele per generation, and thus we expect the frequency of an allele maintained by mutation–selection balance to be small. We therefore approximate the change from selection

We refer to a balance between mutation and selection because in our model the two forces cause changes of opposite sign in the frequency of the allele a.

If q is small, we can approximate $\overline{w}$ by 1.

$$\Delta q = q' - q = \frac{qw_a}{\overline{w}} - q \approx qw_a - q. \qquad (3.69)$$

The change in allele frequency of a due to mutation is approximately μ. Using the parameterization in Table 3.5, we obtain the equation for equilibrium allele frequency under mutation–selection balance by adding μ to the right-hand side of (3.69):

We use the relation $w_a = p(1 - hs) + q(1 - s)$.

$$q[p(1 - hs) + q(1 - s)] - q + \mu = 0. \qquad (3.70)$$

We now look at two cases: when a is a recessive allele with $h = 0$, and when a is a dominant allele with $h = 1$.

If $h = 0$, equation (3.70) becomes

$$q[p + q(1 - s)] - q + \mu = 0. \qquad (3.71)$$

Simplify this to get

Use $p + q = 1$.

$$-q^2 s + \mu = 0, \qquad (3.72)$$

which has the solution

$$q = \sqrt{\frac{\mu}{s}}. \qquad (3.73)$$

If $h = 1$, equation (3.70) becomes

$$q[p(1 - s) + q(1 - s)] - q + \mu = 0. \qquad (3.74)$$

Simplify this to get

Use $p + q = 1$.

$$-qs + \mu = 0, \qquad (3.75)$$

which has the solution

$$q = \frac{\mu}{s}. \tag{3.76}$$

We have found that the frequency of a is given by

$$q \approx \frac{\mu}{s} \tag{3.77}$$

in the case where a is dominant, $h = 1$. The frequency of a is given by

$$q \approx \sqrt{\frac{\mu}{s}} \tag{3.78}$$

in the case where a is recessive, $h = 0$. Thus, mutation can maintain a recessive allele that is selected against at relatively high frequency. Even a lethal recessive will have a frequency of $\sqrt{\mu}$, which for a mutation rate of 10^{-4} is 0.01 and for a mutation rate of 10^{-6} is 0.001. As the frequency of a recessive lethal depends only on the mutation rate, this frequency can be used to provide an indirect estimate of mutation rates. However, it should be kept in mind that this is only the mutation rate to recessive lethals and not an estimate of mutation rates in general.

For a lethal recessive, $h = 0$ and $s = 1$.

3.5 Selection and optimization

One of the ways that ecologists use population genetics and evolutionary theory is to understand and justify optimization arguments. Often ecologists look for a 'fit' between organisms and their environment, assuming that the organism is perfectly adapted to its environment. We are thus led to two questions:

- What does evolution optimize? This question is too general, so we will assume that a population has reached a stable equilibrium, and ask what is optimized at a stable polymorphic equilibrium. (We could also ask what is optimized at a stable fixation equilibrium.)
- How quickly do populations approach an optimum? This question would certainly arise if we wanted to know the response of species to a changing environment. Although in

the models thus far we have assumed a constant environment, we know that environments are continually changing.

Maximization of mean fitness

What is optimized at a stable polymorphic equilibrium? We will show that *mean fitness of the population* is optimized in our two-locus, two-allele model. Start with the formula (3.28) for $\overline{w}$:

Substitute $1 - p$ for q and multiply out to make taking the derivative easier.

$$\overline{w} = p(1 - ps) + q(1 - qt) \tag{3.79}$$
$$= p - p^2s + (1 - p)(1 - (1 - p)t) \tag{3.80}$$
$$= p - p^2s + 1 - p - t + 2pt - p^2t \tag{3.81}$$
$$= -p^2s + 1 - t + 2pt - p^2t. \tag{3.82}$$

To determine the maximum of the mean fitness of the population, as a function of the allele frequency, take the derivative of (3.82) and set it equal to zero:

$$\frac{d\overline{w}}{dp} = -2ps + 2t - 2pt. \tag{3.83}$$

Setting $\frac{d\overline{w}}{dp} = 0$

$$0 = -2ps + 2t - 2pt. \tag{3.84}$$

Solving for p

$$p = \frac{t}{s + t}. \tag{3.85}$$

We recognize this as the allele frequency at a polymorphic equilibrium. Unfortunately, this simple result does not exactly hold for more complex models with more loci; in general, maximization of fitness does not exactly correspond to a stable equilibrium. However, in many cases, the result is approximately true. This population genetics result is used as a justification for assuming optimality in an ecological context, as we will do later in the book.

Rate of change of fitness

How fast can a population respond to changes in its environment? This is an important question, and the answers have many practical implications. Will populations respond quickly enough to global change? We expect predators to evolve to become more efficient at eating their prey, and prey to evolve to become better at escaping their predators. What will be the outcome of this 'evolutionary arms race'?

We use the haploid model because it is easier to analyze. This approach, of using a haploid model to simplify analysis, has been often used in population genetics. For the present question, the answer remains essentially the same if we look at a diploid model, but the analysis is more complex.

Rather than studying the diploid population genetic model we have used thus far, we will present an analysis of a simple model that illustrates the basic ideas, following the presentation in Roughgarden (1979). Instead of using diploid organisms, we will look at haploid organisms, or clones.

As in the rest of this chapter, we will ignore density dependence, so the fitness of a clone i is simply the growth rate of that particular clone, r_i. Each clone will be assumed to be growing exponentially. The population size of clone i will be n_i. If we let $N = \sum n_i$ be the total population size, it makes sense to define the mean fitness of all the clones as

This makes sense because the definition of fitness here is the growth rate.

$$\bar{r} = \frac{\sum r_i n_i}{N}. \tag{3.86}$$

To find the response to selection, we need to find the rate of change of the mean fitness.

We first consider the role of among the growth rates of the different clones. This variability represents the raw material on which selection can act. Without variability, selection cannot act, and there can be no change in the mean fitness. Variability can be quantified as *variance*, and we compute the variance in the growth rate using the standard formulas that can be found in any statistics book:

$$\sigma_r^2 = \frac{\sum n_i (r_i - \bar{r})^2}{N} \tag{3.87}$$

$$= \frac{\sum n_i r_i^2}{N} - \bar{r}^2. \tag{3.88}$$

Use the definition of $\bar{r}$.

The rate of change of mean fitness is

Use $\frac{d}{dt}\frac{u}{v} = (v\frac{du}{dt} - u\frac{dv}{dt})/v^2$.

$$\frac{d\bar{r}}{dt} = \frac{d}{dt}\left[\frac{\sum r_i n_i}{N}\right] \tag{3.89}$$

Use $\frac{dn_i}{dt} = r_i n_i$ and $\frac{dN}{dt} = \bar{r}N$.

$$= \frac{\left(\sum r_i \frac{dn_i}{dt}\right) N - \frac{dN}{dt}\left(\sum r_i n_i\right)}{N^2} \tag{3.90}$$

$$= \frac{\sum r_i^2 n_i}{N} - \frac{\bar{r}N \sum r_i n_i}{N^2} \quad (3.91)$$

Use $\sum r_i n_i = \bar{r}N$ from equation (3.86).

$$= \frac{\sum r_i^2 n_i}{N} - \bar{r}^2 \quad (3.92)$$

Observe that this last expression is simply the variance in the growth rates, (3.88). Thus, we have confirmed our intuition that variability is required for selection to proceed. In fact, the rate of response to selection is proportional to the genetic variance in the population. For this reason, circumstances that reduce the genetic variance of a population are worrisome for the long term preservation of the population.

For a diploid population, a similar result holds: the rate of response to selection depends on the additive genetic variance. For a discussion of additive genetic variance, see Falconer (1989).

3.6 Drift

We now turn to a process that can be very important in removing variability in small populations. By chance, some individuals in a population leave behind no offspring while others leave behind more offspring. This process is called *genetic drift*, and clearly it has its largest effect in small populations. Obviously, eventually drift – also known as random sampling – leads to the elimination of variability, and drift proceeds more quickly in smaller populations. Note that the loss of variability in small populations (fewer than 50) has been raised as an important issue in conservation biology. Before briefly returning to the rate of loss of variability caused by drift, we will look at the role of drift in another way, as a force causing changes in the alleles present in populations.

Faced with the large number of alleles present at many loci in most populations, some geneticists (notably Kimura) emphasized the importance of a balance between mutation and random forces in maintaining variability and in leading to the kinds of substitutions detectable by looking at sequences of DNA or amino acids in proteins. The idea was that mutations would produce new alleles, and that random sampling effects could lead to the increase in frequency of these alleles.

Drift would eliminate most new alleles, but some alleles would increase in frequency.

As an example consider a population of a diploid organism with a single individual. Assume initially that the individual (and population) is homozygous *AA*. If the mutation rate is μ per generation

Reproduction in this population occurs by randomly choosing one of the two possible alleles, and then repeating the process. There are organisms, including some snails, that can self-fertilize, leading to exactly this kind of genetic process.

per allele, then on average, a new allele would be expected to arise by mutation every $\frac{1}{2\mu}$ generations. Denote this allele as a. If the only force operating is random sampling then a single *Aa* individual leaves behind a *aa* individual with probability 1/4, an *Aa* with probability 1/2, and an *AA* with probability 1/4. Reasoning by common sense, we conclude that the probability that this new mutation becomes 'fixed' in the population is 1/2. Thus, making use of the idea that the time between mutations is large relative to the time for an allele to become fixed, we conclude that the rate at which neutral alleles are substituted in a population fixed at size 1 is is $\frac{1}{4\mu}$. We can then guess the general rate of substitution $\frac{1}{4N\mu}$, where N is the *effective population size*, which was one in the simple example. A random mating population with equal numbers of males and females has an effective population size equal to the population size; deviations from random mating reduce the effective population size because then drift operates faster.

In this view, variability at the molecular level is seen as a balance between drift and mutation. Note that by adjusting population size and mutation rate the model can yield a wide range of levels of variability. Moreover, mutation rate and population size are not easy parameters to measure. Thus, it can be very difficult to reject the hypothesis that genetic variability in populations at the molecular is caused by a balance between drift and mutation.

We have also essentially computed the rate at which drift eliminates variability. The argument above implies that in a population of size N, a fraction $\frac{1}{2N}$ of the variability is lost each generation. Thus, a population of size 50 – the minimum safe size suggested by some experts in conservation genetics – loses 1% of its genetic variability by drift each generation. A smaller population would lose variability at a faster rate, without any balancing increase in the rate at which genetic variability was added.

This rule of 50 has been criticized as being too simplistic by some scientists.

3.7 Ecology and evolution

The work we have discussed so far is important for understanding how genetics relates to ecology, but there is an important additional case that we now consider. Ecologists are interested in interactions between individuals, so the fitness of an individual will not be constant. Instead it will also depend on the other members of the population with which it interacts. Thus, we will look at the role of fitnesses that depend on the frequencies of different types in the population, which is very important for understanding the evolution of behavior.

So far we have assumed that fitnesses are constant, but now we will look at cases where there is frequency dependence in the fitnesses.

Evolutionarily stable strategies

In this approach we ignore explicit genetics, because we focus on the role played by frequency dependence. We follow the approach outlined in Maynard Smith (1982, 1989) quite closely. Consider the fact that in many animals, such as the East African oryx, fights are determined by 'ritual' behavior: at some point the loser in the fight withdraws. Why doesn't the loser just keep fighting until one animal is injured or dies? Arguments based on 'the good of the species' are not acceptable, as geneticists have shown that there is no known evolutionary mechanism that will produce behavior that is 'good for the species'.

Assume that there are two kinds of animals in the population: Hawks and Doves. Hawks always escalate fights until one animal is injured, while Doves always retreat if a fight is escalated. We will assume that in encounters between two Doves, or between two Hawks, the winner is determined randomly. We assume that in a fight between two Hawks (the only one that can lead to an injury), there is a cost, C (in terms of future reproduction), to the injured animal. Winning the contest in any case leads to a gain of V.

A 'currency' for these costs and gains that would make biological sense is reproductive value, which we defined in the previous chapter.

We will attempt to determine the *evolutionarily stable strategy* (or ESS) in this case. What is an ESS? It is a strategy that has a higher expected payoff (for an individual) than any other strategy when the strategies it is played against are those currently in the population. This is easiest to understand by example, so let us return to the Hawks and Doves.

We first list the expected payoffs (the gains and costs mentioned above) in Table 3.6. First consider the case where $V > C$. Then the strategy consisting of always being a Hawk is an ESS, which

We ignore the case of two Doves because when Doves are rare virtually every encounter a Dove has will be with a Hawk.

TABLE 3.6. Payoff matrix for Hawk and Dove game. The gains (which are positive) and costs (which are negative) from a single interaction between two animals are listed.

	opponent	
Payoff to:	Hawk	Dove
Hawk	$(V - C)/2$	V
Dove	0	$V/2$

we can see as follows. When Dove is rare we can ignore contests involving two Doves. The expected payoff to a Hawk is the payoff of a contest between two Hawks, which is positive. The expected payoff to a Dove is zero. Thus in this case Hawk has a higher 'fitness'. Here, the cost of injury is lower than the reward of victory.

Next consider the case where $V < C$. Is the pure Hawk strategy an ESS? Again we ignore contests involving two Doves. The expected payoff to a Hawk is the payoff of a contest between two Hawks, which is now negative. This is less than the expected payoff to a Dove when facing a Hawk, which is zero. Thus we expect the Dove strategy to increase when rare. If the cost of injury is high relative to the reward, a pure fighting strategy is not an ESS.

We refer to the increase of the Dove strategy, rather than the increase of Doves, because in this model we are only interested in the frequency of the strategy and do not specify how the strategy is determined. The strategy could be determined by genetics, or could be a behavioral choice.

Is the pure Dove strategy an ESS when $V < C$? If Hawks are rare, the expected payoff to a Hawk from an encounter (necessarily with a Dove) is V, which is greater than the expected payoff to a Dove encountering another Dove, which is $V/2$. Thus the pure Dove strategy is never an ESS.

Even if the payoff to both strategies is the same, there will be an ESS only if the payoffs to both strategies goes down as the frequencies change – the payoffs instead could be equal, but at a minimum.

We guess that if $V < C$ a mixed strategy is an ESS. We will look for a *mixed* strategy that leads to the same payoff to Hawk and Dove. If such a strategy exists, there will be no tendency for Hawk or Dove to increase when this strategy is used. Assume that Hawk is played with probability P, and Dove with probability 1-P. (This can result from each animal playing a mixed strategy, or from a mixture of animals.) We do not distinguish between a hypothetical population where a certain fraction of animals are Hawks and a certain fraction are Doves, and another hypothetical population where each animal uses each strategy the corresponding fraction of the time. We do not specify the genetic basis for the strategies.

The expected payoff to a Hawk is found as the probability that the opponent is a Hawk times the payoff when the opponent is a Hawk plus the probability that the opponent is a Dove times the payoff when the opponent is a Dove. Thus the expected payoff to a Hawk is

$$P(V - C)/2 + (1 - P)V. \tag{3.93}$$

Similarly, the expected payoff to a Dove is

$$(0)P + (1 - P)(V/2). \tag{3.94}$$

We equate the payoffs to Dove and to Hawk, because if the two payoffs are equal we expect the frequencies of the two strategies in the population to remain constant:

$$P(V - C)/2 + (1 - P)V = (1 - P)(V/2) \tag{3.95}$$

Solving this equation for P, we find

$$P = V/C. \tag{3.96}$$

We have now found the mixed ESS. This says that the fraction of animals that are Hawks is given by V/C, so that if the payoff from each encounter is small relative to the cost from injury, a large fraction of the animals should be Doves.

To be certain that this mixed strategy is truly an ESS and not a minimum in terms of payoffs, we should also demonstrate that a small change in the frequency of Hawks or Doves does not lead to an increase in the payoffs, but to a decrease in the payoffs.

This kind of argument can be used to understand the evolution of many behavioral features, ranging from dispersal rates to care of young. A summary of the procedure for finding an ESS is given in Box 3.4.

Problems

1. In a population you find that the frequency of A and a are 0.95 and 0.05, respectively. Assume that the population is at equilibrium.
   (a) Determine parameter combinations that could lead to this observation on the basis of heterozygote advantage. In other words, pick values for w_{AA}, w_{Aa}, and w_{aa}. You might find it handy to use the parameterization where we call these values $1 - s$, 1, and $1 - t$. In

Box 3.4. Finding an evolutionarily stable strategy.

To find an ESS for the game between the two strategies A and B with the payoff matrix

	opponent	
Payoff to:	A	B
A	α	β
B	δ	γ

we use the following series of steps. (For simplicity, assume that there is no more than one mixed strategy yielding equal payoffs to both types. If there is, an extension of this procedure will determine the ESS's.)

- First determine if either (or both) pure strategies are ESS's. To determine if pure B is an ESS, compute the payoff to type A when rare, β, and compare it to the payoff to B when common, γ. If the payoff to B is higher then it is an ESS. Do the same with pure A.
- Next, determine if there are one or more mixed strategies with equal payoffs. Set the frequency of strategy A to p (and that of B to $1 - p$). Set the payoff to A, $p\alpha + (1 - p)\beta$, equal to the payoff to B, $p\delta + (1 - p)\gamma$, and solve for p, restricting p to values between 0 and 1.
- If just one pure strategy is an ESS and there are no other ESS's, then that pure strategy is the ESS.
- If neither pure strategy is an ESS, then the mixed strategy with A used a fraction p of the time. (Since we assume a single mixed strategy yielding equal payoffs, the fact that neither pure strategy is an ESS implies that payoffs are maximized when payoffs are equal.)
- If both pure strategies are ESS's, then the outcome is determined by initial conditions – the system will end up with all type A if the initial frequency of A is higher than p. Otherwise the system will end up with all B.

this case you need to select a value for s and a value for t. The choices are not unique; to do this you may need to first just pick a value for s and then find t.

(b) Determine parameter combinations that could lead to this observation on the basis of mutation–selection balance with A recessive. Here you need to find μ and s, noting that s is not the same here as s in part a.

(c) Determine parameter combinations that could lead to this observation on the basis of mutation–selection balance with A dominant.

(d) Discuss briefly how, if at all, you might distinguish among these alternative explanations.

2. Discuss briefly (one half page at most) how you might distinguish between mutation and drift on the one hand and selection on the other hand as an explanation for an observed polymorphism of the form in Problem 3.1.

3. In many areas of the world where malaria was common, there are alleles that are lethal when homozygous but are advantageous when heterozygous by conferring resistance to malaria. These alleles can reach a frequency as high as 40%. If we assume that the fitness of hh individuals is zero, and the frequency of h is 0.1, what are the relative fitnesses of HH and Hh individuals? What would the fitnesses be if the frequency of h were 0.25 or 0.5?

4. Although the frequency of dominant alleles causing industrial melanism reached a very high level relatively quickly in England, the frequency of (recessive) alleles causing the light forms typically did not drop below 10%. What do you know about the dynamics of selection (not depending on frequency-dependent selection) that could explain this observation?

5. Assume that you are trying to detect selection by seeing if adult genotype frequencies fail to fit Hardy–Weinberg proportions. Assume that the relative fitness of the two homozygotes is 0.9 and that of the heterozygote is 1.0. The statistical test you use to determine if the deviations from Hardy–

Weinberg are significant is to assume that the expected frequencies of the three genotypes are p^2, $2pq$, q^2 and compare the expected numbers to the observed numbers. Do the comparison by computing the quantity

$$\sum \frac{(\text{observed} - \text{expected})^2}{\text{expected}}$$

where the sum is over the three genotypes, and the observed and expected are the observed and expected **numbers**, not frequencies, of each of the genotypes. This quantity is a chi-square with 1 degree of freedom, and the result is significant at the 0.05 level if the quantity is greater than 3.54. Assume that the observed frequencies are exactly those predicted by selection at equilibrium. Then the quantity will depend on the number in the sample/population. How big must this number be so that the quantity you calculate is greater than 3.54, that is, so that you detect selection? Discuss the implications of your answer.

6. Use the material in the chapter to explain why small populations may lack the ability to respond to environmental changes. (What is the level of genetic variability, and why does it matter?)

7. Individuals are of two types, A and B. They interact in random pairs with payoffs as shown. What are the evolutionarily stable state(s) of the population in each case?

(a)

	opponent	
Payoff to:	A	B
A	6	4
B	1	5

(b)

	opponent	
Payoff to:	A	B
A	3	4
B	5	2

8. This problem is a modification of a problem in Maynard Smith (1989). Explaining the evolution of alarm calls can be difficult, because the bird sounding the alarm exposes itself to a higher risk of predation. However, we can use the

ESS approach to look at this problem. We will idealize the situation so that all interactions are between pairs, and there are only two possible behaviors: sounding an alarm, or not sounding an alarm. You will determine the ESS that yields the highest fitness in this case. If a bird sounds an alarm, it has a 50% chance of survival. If a bird does not sound an alarm, but is with a bird that does sound an alarm, then the bird that does not sound the alarm has a 100% chance of survival. If neither bird sounds an alarm then each bird has a 25% chance of survival.

(a) Set up a payoff matrix for this situation, focusing on individual incidents.

(b) Determine the ESS.

(c) Determine how the ESS would be changed if the specific probabilities of survival for each possible set were changed, but the rank order of survival probabilities remained the same (not sounding an alarm while with a bird that sounds an alarm has the highest survival and neither bird sounding an alarm has the lowest survival).

Suggestions for further reading

In this chapter, we have just begun to touch upon some of the extensive issues of population genetics. Good population genetics textbooks include Crow (1986), Hartl and Clark (1988), and Maynard Smith (1989). A standard textbook on evolutionary biology is Futuyma (1986). Falconer (1989) is a standard reference on quantitative genetics.

The case of industrial melanism is discussed in Kettlewell (1956). Endler's (1986) book, *Natural Selection in the Wild*, is a detailed review of attempts to detect selection acting in natural populations. This book is also useful for a more general consideration of the problem of determining a mechanism producing an observed change in a biological population.

The neutral theory was first proposed by Kimura (1968), and was given a book-length treatment in Kimura (1983). A selectionist explanation for the maintenance of variability at the molecular level is presented in the book by Gillespie (1991).

The book *Evolution and the Theory of Games* by Maynard Smith (1982) gives a thorough treatment of ESS's. The book *Ecological Genetics* edited by Real (1994) has a series of overviews on topics in this important area.

4

Density-Dependent Population Growth

We now return to strictly ecological questions. We have remarked earlier that our models and common sense imply that exponential growth cannot continue forever. As illustrated by the growth of collared doves in Great Britain (Figure 2.3), growth of natural populations cannot be exponential forever, and eventually approaches zero. Similarly, the number of sheep in Tasmania increased after introduction in a fashion that could be considered exponential, but then the numbers reached an approximate equilibrium (Figure 4.1). In this chapter we return to the fundamental question of the causes and consequences of regulation of population growth.

4.1 Hypotheses for population regulation

What regulates the growth of the sheep population in Tasmania, or the growth of other populations? Many hypotheses have been proposed for the causes of regulation of populations:

- Populations are limited by density-independent factors such as changes in the weather.
- Populations are limited by their food supply.

Can you think of other factors that might regulate population growth?

How useful is it to think of a single factor, as opposed to a combination of factors, as the cause of population regulation?

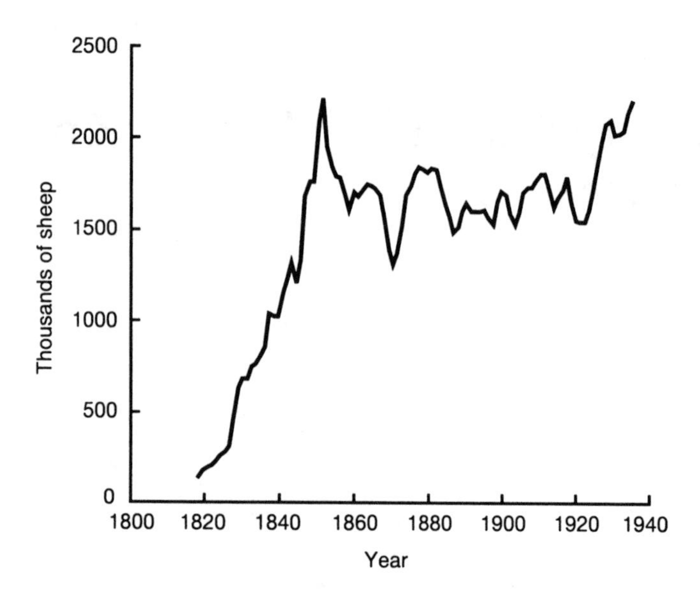

FIGURE 4.1. Dynamics of sheep numbers in Tasmania after introduction (data from Davidson, 1938).

- Populations regulate themselves through mechanisms such as territoriality or cannibalism.
- Populations are regulated through competition.
- Populations are regulated by predators.
- Populations are regulated by parasites or diseases.

We will explore some of these hypotheses in later chapters. Here we begin a discussion of the implications of regulation by limitations in the food supply, or more generally, density-dependent factors that operate within a single species. One of our eventual goals is to elucidate information about the causes of population regulation by examining the consequences of different forms of population regulation. Obviously, any of these factors would limit growth, so the important question is what in nature is actually the limiting factor for a particular population.

Density dependence here refers to processes whose effect changes as the number of individuals within the population changes. Thus regulation by food limitations is one example of a density-dependent factor.

How might you experimentally determine the regulating factor in a natural population?

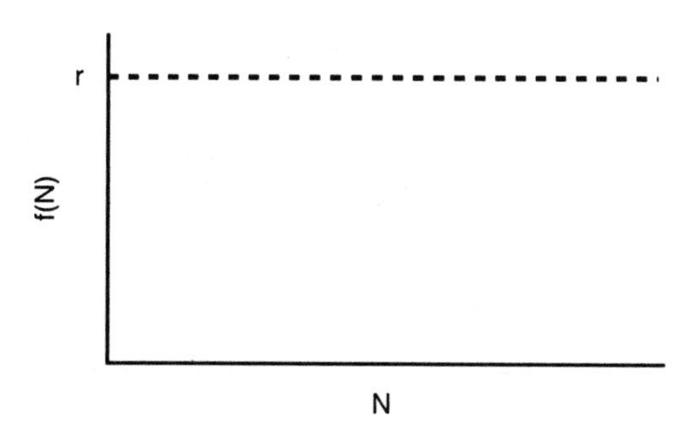

FIGURE 4.2. Per capita growth rate in a model without density dependence

4.2 Logistic model

Regulation by limitations in the food supply can be a difficult question to address through field studies, although much progress has been made recently. Laboratory studies by Gause (1934, 1935) and others demonstrated early in the twentieth century the dynamics of populations limited by their food supply. We will explore models that demonstrate the dynamics expected.

The basic model we will examine takes the form

$$\frac{dN}{dt} = Nf(N), \tag{4.1}$$

We ignore here the consequences of age structure. Coupling age structure with density dependence is in fact a difficult problem that is the subject of much current research.

where $f(N)$ is the per capita growth rate. By writing the equation this way, we emphasize that the per capita growth rate may depend on the number of organisms in the population, thus providing one explanation for the cessation of exponential growth. In the simplest models that we explored earlier, the per capita growth rate $f(N)$ was a constant r and did not depend on the population density N, as shown in Figure 4.2.

Here, the parameter r couples both births and deaths, which in fact causes some problems with biologically interpreting the parameter K in the logistic model.

We now change the form of the function $f(N)$ to include the effects of density dependence. What is the simplest function that changes with density so the per capita growth rate goes down as the density goes up? It is a straight line, as illustrated in Figure 4.3. We assume that when the population density is very small, the per capita growth rate is given by r, the 'intrinsic rate of increase'. We denote by K the value of the population density at which the per capita growth rate is zero. This is known as the *carrying capacity*. Note that we have not specified the biological mechanism

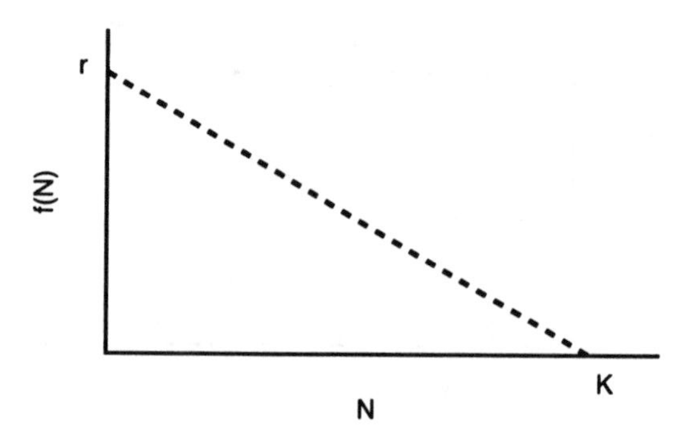

FIGURE 4.3. Per capita growth rate in a model with density dependence.

responsible for this dependence on density. What is the equation describing this line? We write it as

$$f(N) = r(1 - N/K). \tag{4.2}$$

It is easy to see that this is the equation of a straight line and that it satisfies the two conditions that $f(0) = r$ and that $f(K) = 0$. This model is known as the *logistic model.*

Explicit solution of the logistic model

This model is simple enough that we can find the explicit solution and determine the density N as a function of time. The method we use cannot be applied to more complex models, but is useful here. We begin with the model

We are interested in the explicit solution because we compare it to data below. The answer here is more important than the technique, so if you have trouble following the steps, just concentrate on the final answer, equation (4.12).

$$\frac{dN}{dt} = rN(1 - N/K). \tag{4.3}$$

We then separate variables by writing all the terms with N on one side of the equation and all the terms with t on the other side:

$$\frac{dN}{N(1 - N/K)} = rdt. \tag{4.4}$$

We then integrate both sides of the equation from $t = 0$ to $t = T$:

$$\int_{N(0)}^{N(T)} \frac{dN}{N(1 - N/K)} = \int_0^T rdt. \tag{4.5}$$

To integrate the left-hand side of this equation, we use the technique of integration using partial fractions, first writing

Your calculus textbook has a section on integration using partial fractions.

Here we first write $\frac{1}{N(1-N/K)} = \frac{a}{N} + \frac{b}{1-N/K}$ and then determine a and b.

$$\frac{1}{N(1-N/K)} = \frac{1}{N} + \frac{1/K}{1-N/K}. \tag{4.6}$$

Then

$\int \frac{1/K}{1-N/K} = -\ln(1-N/K)$

$$\int_{N(0)}^{N(T)} \frac{1}{N} + \frac{1/K}{1-N/K} dN = \left[\ln(N) - \ln(1-N/K)\right]_{N(0)}^{N(T)} \tag{4.7}$$

$$= \ln(N(T)) - \ln(1-N(T)/K)$$
$$- \ln(N(0)) + \ln(1-N(0)/K). \tag{4.8}$$

The right-hand side of (4.5) is simply

$$\int_0^T r dt = rT. \tag{4.9}$$

Combining these last two equations, we see that

$$\ln(N(T)) - \ln(1-N(T)/K) - \ln(N(0)) + \ln(1-N(0)/K) = rt. \tag{4.10}$$

Taking the exponential of both sides, we find that

Recall that $e^{a+b} = e^a e^b$ and that $e^{\ln(a)} = a$.

$$\frac{N(T)(1-N(0)/K)}{(1-N(T)/K)N(0)} = e^{rt}. \tag{4.11}$$

Solving this equation for $N(T)$, we find

$$N(T) = \frac{N(0)e^{rT}}{1+N(0)(e^{rT}-1)/K}. \tag{4.12}$$

What does this equation imply that $N(T)$ is (approximately) when T and $N(O)$ are small? What does $N(T)$ approach when T is large?

This solution is graphed in Figure 4.4.

Comparison with data

The very general result that we have just obtained, that the population first grows more rapidly and then approaches an equilibrium, is exactly what we have seen for collared doves (Figure 2.3) and sheep in Tasmania (Figure 4.1). We can now ask how exact are the predictions – is just the general feature of approach to an equilibrium correct, or does population growth really follow a logistic curve?

We can turn to simple laboratory experiments on microorganisms as studied by Gause (1934, 1935). As illustrated in Figure 1.1, the dynamics of the microorganism *Paramecium aurelia* show

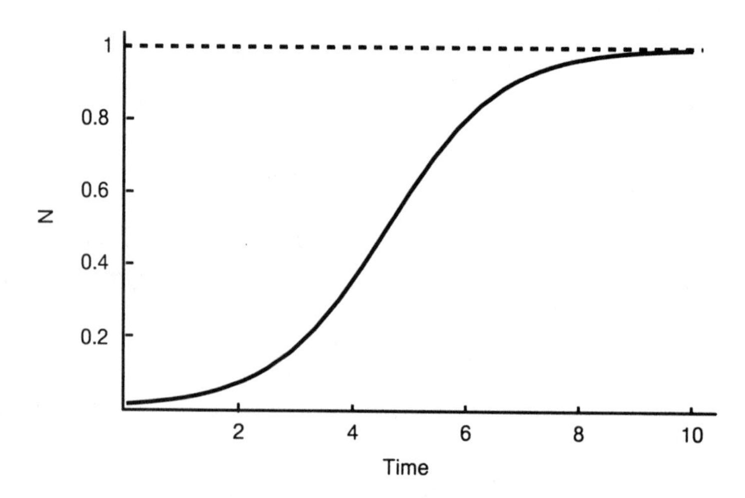

FIGURE 4.4. Explicit solution of the logistic model with $r = 1, K = 1$, and $N(0) = 0.01$.

what appears to be logistic growth during the first 10 days of the experiment. However, the population then declines and appears to approach a second equilibrium phase. Thus, even in the simplest setting, our model can be an oversimplification. We will now begin asking how to modify the simple model to make it more realistic.

Why do we say 'describe', rather than 'predict', population growth?

We can also ask how good the quantitative fit of the model is to data; perhaps the logistic model describes the initial phase of population growth well. One way to do this is to find the best fit of the model to experimental data and ask how good the fit is. One way to find the best fit is to calculate the parameter values that minimize the sum of the squares of the deviations of the model from the experimental data points, where the sum is taken over the experimental data points. One example of this kind of approach is given in Figure 4.5, where the numerical fit is quite good. One should, however, exercise caution in taking fits like these as proof that the logistic model is a good description of population dynamics. Not only do we already have the example given in Figure 1.1, but we can use the following example as another cautionary tale.

A celebrated example demonstrating the quantitative failure of the logistic model is the growth of the human population in the United States (Figure 4.6). Pearl et al. (1940) fit a logistic curve to

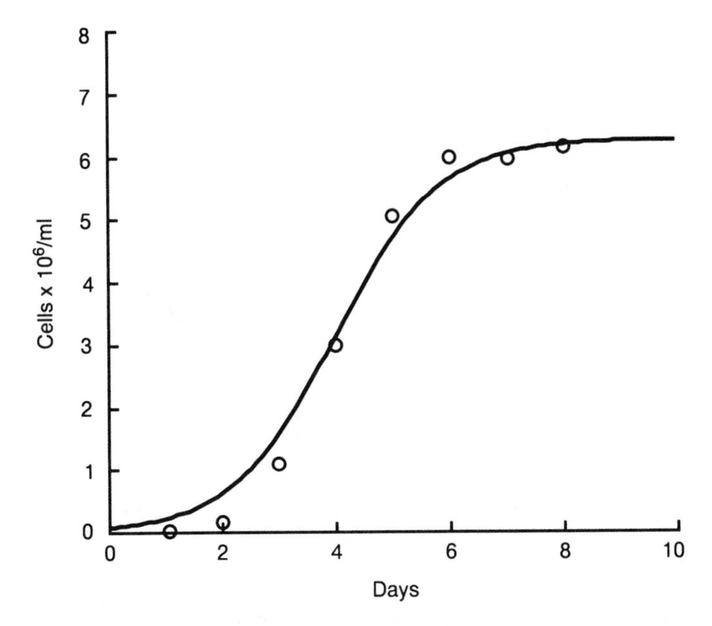

FIGURE 4.5. Plot of numbers of *Escherchia coli* versus time from an experiment of McKendrick and Pai (1911). The best fit of the logistic model is also drawn.

population censuses in the United States through 1940. Although the logistic model fit the data until 1940 very well, the fitted curve predicted that the population of the United States would level off, when in fact it has continued to rise. Thus, we are led to the conclusion that we should not try to make too much of the quantitative aspects of the logistic model.

Why do you think the logistic model fit so well, but did not predict future population levels?

'Quantitative' here refers to numbers, while 'qualitative' would mean more general features, such as the prediction that the population reaches an equilibrium.

Equilibrium analysis

We now outline how a qualitative analysis of the logistic model would proceed. Although we have solved the logistic model exactly, there are several reasons to go through this analysis. First, more complex models cannot be solved exactly, so it is useful to learn this technique. By beginning with the logistic model, which we already understand, we can see better how this technique works. Second, unlike the quantitative predictions of the logistic model, which we have indicated are unreliable, the qualitative predictions are *robust*. By robust, we mean that they hold even if we make changes in our model.

Robust models are very important in ecology because the models are typically quite crude – many simplifying and perhaps inaccurate assumptions are usually made.

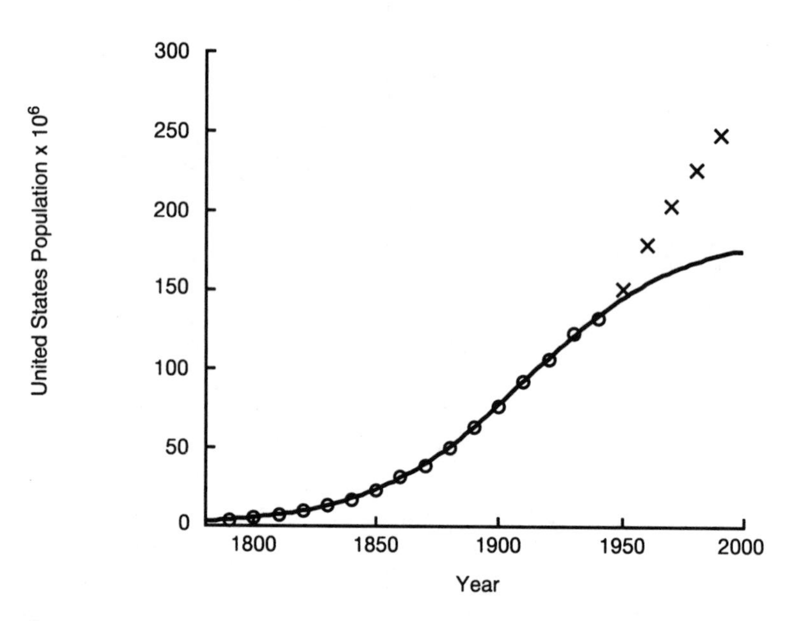

FIGURE 4.6. United States census figures from 1790 until the present. The best fit of the logistic model to the data from 1790 through 1940 (indicated on the graph by circles with the fit drawn as the curve) is astoundingly good, but the fit since that time is very poor (the data given by X's). This example, based on the fit to the data until 1940 by Pearl et al. (1940), is discussed in Hutchinson (1978).

We will outline the analysis as a series of steps:

- Determine the values of the population density, $\hat{N}$, which are equilibria. Set

$$\frac{dN}{dt} = 0 \tag{4.13}$$

to obtain

$$r\hat{N}(1 - \hat{N}/K) = 0 \tag{4.14}$$

which has the solutions

$$\hat{N} = 0 \text{ and } \hat{N} = K. \tag{4.15}$$

The details of the development of this procedure are not absolutely necessary to understand what follows. If particular steps are not clear, keep reading. The concepts and the summary in the box below are important.

- Determine the behavior of solutions near equilibrium points. Near $\hat{N} = 0$, we see that $\frac{N}{K}$ is much smaller than 1. Thus, we neglect the term $\frac{N}{K}$ in (4.3) so

$$\frac{dN}{dt} \approx rN. \tag{4.16}$$

- Because $r > 0$, and the solution of (4.16) is

$$N(t) = N(0)e^{rt}, \tag{4.17}$$

we conclude that solutions grow exponentially when N is small. This is confirmed by looking at the exact solution in Figure 4.4.

The very beginning of population growth in Figure 4.5, the laboratory population of E. coli, also appears to be exponential.

We now indicate a systematic way to arrive at this conclusion. We let n represent the deviation from the equilibrium, so

$$N = \hat{N} + n. \tag{4.18}$$

Then, in general, we are interested in finding how n changes with time. We observe that since $\hat{N}$ is a constant

$$\frac{dN}{dt} = \frac{dn}{dt}. \tag{4.19}$$

So

$$\frac{dn}{dt} = F(N) \tag{4.20}$$

where

$$F(N) = rN(1 - N/K). \tag{4.21}$$

We need to approximate $F(N)$ near the equilibrium, $\hat{N}$. We use a Taylor series (Box 3.2) to see that

$$F(\hat{N} + n) \approx F(\hat{N}) + n \left.\frac{dF}{dN}\right|_{N=\hat{N}}. \tag{4.22}$$

We note that since $\hat{N}$ is an equilibrium, $F(\hat{N}) = 0$. Thus we conclude that:

$$\frac{dn}{dt} \approx n \left.\frac{dF}{dN}\right|_{N=\hat{N}}. \tag{4.23}$$

In the logistic model with

$$F(N) = rN(1 - N/K) = rN - rN^2/K \tag{4.24}$$

we compute

$$\frac{dF}{dN} = r - 2rN/K. \tag{4.25}$$

Near the equilibrium $\hat{N} = 0$,

$$\frac{dn}{dt} \approx n \left(r - 2rN/K\right)|_{N=0} \tag{4.26}$$

$$= rn. \tag{4.27}$$

We already see that some results about stability are robust, because they depend on the sign of the derivative dF/dN at equilibrium, and this sign will not be changed by small changes in the model. We observe that an equilibrium of the continuous time model will be stable (approached from nearby population levels) if increasing the density reduces the growth rate – that is, if there is negative density dependence.

This is the same conclusion we reached earlier.

Near the equilibrium $\hat{N} = K$,

$$\frac{dn}{dt} \approx \left((r - 2rN/K)|_{N=K}\right) n \tag{4.28}$$

$$= (r - 2r)n \tag{4.29}$$

$$= -rn. \tag{4.30}$$

Thus, if n represents the deviation from the equilibrium $\hat{N} = K$, we conclude that

$$n(t) = n(0)e^{-rt}. \tag{4.31}$$

Hence, solutions approach the equilibrium $\hat{N} = K$. The procedure we have gone through is outlined for general equations in Box 4.1.

The form of equation (4.31) indicates again the fact that the exact form of the per capita growth rate chosen for the logistic is not necessarily correct. There is no biological reason that the rate of return to the equilibrium $\hat{N} = K$ should be the same as the rate of exponential growth when the population is small. However, this is inherent in the logistic model. On the other hand, the basic conclusions from the qualitative analysis we have just performed do not depend on the exact form of the logistic. That the equilibrium with $\hat{N} = 0$ is unstable, and that the equilibrium with $\hat{N} = K$ is stable (where K is the population density at which the per capita growth rate is zero), is a feature of many similar models.

The θ-model described in Problem 4.1 is an example of one alternate model.

Graphical approach

One more way to look at the dynamics of a single species is to use a graphical approach. This approach has the advantage of displaying the dynamics for all values of the population density and of displaying easily the effects of small changes in the model.

Box 4.1. Qualitative analysis of a model with a single differential equation.

We consider a model of the form

$$dN/dt = F(N).$$

The first step in the analysis is to determine the equilibria. Do this by setting $dN/dt = 0$ to obtain an equation for $\hat{N}$:

$$0 = F(\hat{N}).$$

Then solve this equation for $\hat{N}$. Note that this may be impossible to do for some functions F.

The next step is to determine the stability of these equilibria by approximating F. We define the deviation from equilibrium, n, by letting $N = \hat{N} + n$ and compute

$$dn/dt \approx \lambda n,$$

where

$$\lambda = \left.\frac{dF}{dN}\right|_{N=\hat{N}}.$$

- The equilibrium is stable, and is approached if the system starts nearby, if λ for that equilibrium is negative.
- The equilibrium is unstable, and the system moves away from the equilibrium if the system starts nearby, if λ for that equilibrium is positive.

The rate of return to the equilibrium, or the rate at which the system moves away from the equilibrium, is determined by λ.

We graph the rate of change of the population, $\frac{dN}{dt}$, against the population size, N (Figure 4.7). From the figure we can easily find equilibria. From the figure we can also determine whether the population is increasing or decreasing for a given population size. This latter information lets us quickly determine stability, confirming our analytical results and further illustrating that small changes in the model do not lead to qualitative changes in behavior.

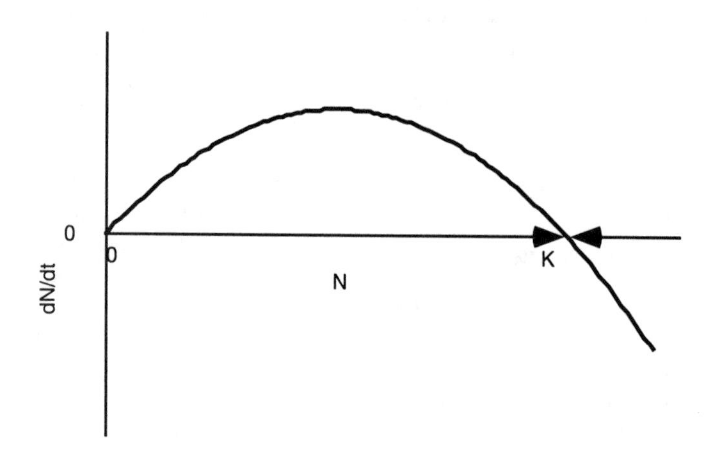

FIGURE 4.7. Plot of rate of change of population size against population size for the logistic model. When $\frac{dN}{dt} > 0$, N increases; when $\frac{dN}{dt} < 0$, N decreases. Along the N axis the direction of population change is indicated by the arrows. From the figure we easily see that the equilibrium $N = 0$ is unstable and the equilibrium $N = K$ is stable.

4.3 Lag time and density dependence

Testing the logistic model is quite difficult, and as we have already suggested, a quantitative test is probably not appropriate. Additionally, if we concentrate on the qualitative features rather than the quantitative ones, the prediction that there is a stable equilibrium is not easily proved or disproved for many examples, like the Tasmanian sheep illustrated in Figure 4.2.

However, a number of natural (and laboratory) populations such as lemmings (Shelford, 1943) show cyclic behavior, a qualitative behavior that cannot be explained by the logistic model. For example, the rotifer populations graphed in Figure 1.1 show strong evidence of cyclic dynamics. We will indicate how this behavior can be explained in general.

Cyclic behavior here means a population whose numbers increase and decrease in a relatively regular fashion.

An analogy

The temperature in the room you are sitting in (unless you are outdoors) is probably controlled by a thermostat. If the temperature gets too low, the heat immediately goes on; if it gets too high the heat goes off. The temperature remains relatively constant. However, suppose that instead someone checked the tempera-

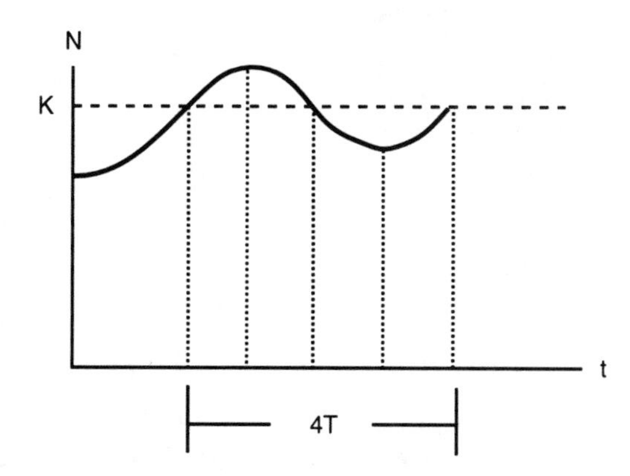

FIGURE 4.8. Qualitative behavior of the delayed logistic model.

ture only once per hour, and either turned off the heat if it was too hot, or turned on the heat if it was too cold. The temperature would almost certainly alternate between too hot and too cold. If the temperature was checked once a week instead, the fluctuations in the temperature would be much more severe. The fact that delayed regulation leads to oscillations is well known in engineering. We will indicate how this principle has been used as a general explanation for oscillations in population biology.

Logistic growth with lags

The simplest model incorporating delayed regulation is the equation

$$dN/dt = rN(t)\left[1 - \frac{N(t-T)}{K}\right], \tag{4.32}$$

originally introduced by Hutchinson (1948) and first carefully analyzed by Wangersky and Cunningham (1957). From the qualitative argument we have just made, we conclude that if the delay is long enough, then the model will exhibit oscillations.

We will now go through a qualitative argument leading to the conclusion that the period of the oscillations will be roughly $4T$, as indicated in Figure 4.8. Assume that first the population is below the carrying capacity. It will increase, and the per capita growth will be positive. The per capita growth rate will remain positive until T time units after the population rises above K. At

TABLE 4.1. Cyclic behavior of the delayed logistic model.

rT	N_{max}/N_{min}	cycle period
1.57 or less	1.0	no cycles
1.6	2.56	4.03
1.9	22.2	4.29
2.5	2930	5.36

this point the population will begin to decline. It is reasonable to assume that it will take about T time units to fall back to the carrying capacity. It will continue to decline for another T time units, and then start to rise again. Similarly, we conjecture that it will take about T time units for the population to rise again to the carrying capacity. Thus completion of the cycle takes roughly $4T$ time units.

Note also that in determining stability, we conjecture that the model is unstable if the delay is long enough. But the 'size' of T depends on the units we choose for time. The condition for stability should be independent of the units we use for measurements. Similarly, if the growth rate of the population is high enough, even a short delay may be enough to cause instability. Observe that the units of T are time and the units of r are per time (time^{-1}). Thus the quantity rT has no units; it is dimensionless. It is the size of rT that determines stability of the equilibrium point $\hat{N} = K$ in the delayed logistic model. The exact stability results are as indicated in Table 4.1. The general argument about nondimensional parameters is important, and too little used in ecology. The concept is summarized in Box 4.2.

Nicholson's blowflies

There have been a number of classical laboratory experiments that have been the subject of many modeling studies. One of the prime examples is Nicholson's (1957) blowfly experiment, which examined the dynamics of *Lucilia cuprina* over a long time. May (1975) fit the delayed logistic model to this data.

We use the ideas in Box 4.2.

The logic of May's approach is as follows. There is only one nondimensional parameter combination, namely rT. The combi-

Box 4.2. The role of nondimensional parameters.

In analyzing models we look for parameters or combinations of parameters that determine stability. One important concept that arises is that of *nondimensional* parameters: parameters without units. We could measure biomass of a population in kilograms, pounds, tons, yielding different values for a carrying capacity, since carrying capacity and the population have the same units. But the stability of the model and of the population obviously cannot change as we change the units. Similarly, the units of intrinsic rate of increase are inverse time units, and the stability of the population cannot depend on whether time is measured in weeks, years, or centuries. *Our conclusion is that stability can only depend on nondimensional groups of parameters, combinations for which all the units cancel.* For example, in the delay logistic model, the only nondimensional combination possible is the time delay multiplied by the intrinsic growth rate, and this is the combination that determines stability. One can often identify the stability-determining groups of parameters for a system by simply looking for nondimensional combinations.

nation rT is adjusted to fit the experimental curve. The parameter r is estimated from the life table data, allowing T to be estimated. This value of T is then compared to the observed egg-to-adult time at the experimental temperature.

The best fit to the data is obtained with $rT \approx 2.1$. This leads to a calculated value of $T \approx 9$ days, while the actual value of T is 11 to 14 days, which is a very good fit. This, however, cannot be taken as proof that the model is correct. In fact, some details of the dynamics are clearly not explained by this model, namely the tendency for the data to exhibit a double peak.

Explain why we have not proven that this model is correct.

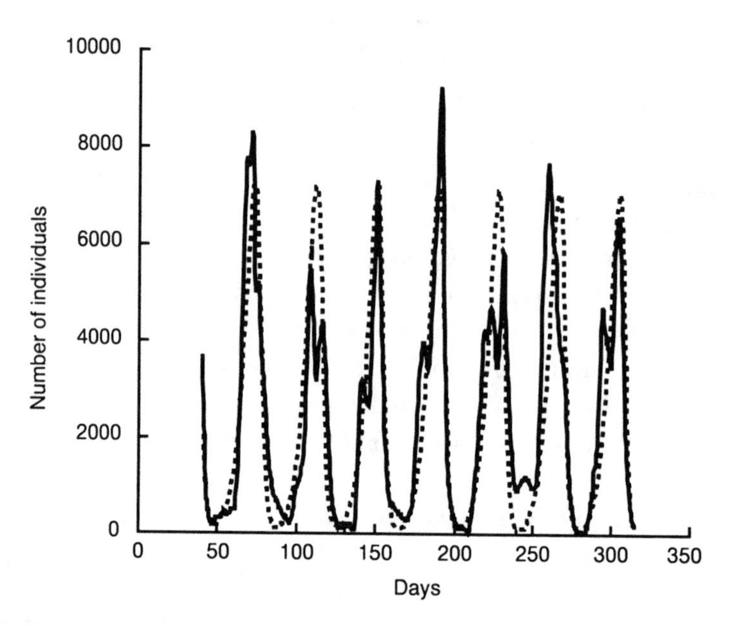

FIGURE 4.9. Plot of numbers of blowflies *Lucilia cuprina* and best fit from the delayed logistic versus time. The data (thick line) are from Nicholson (1957), and the fit (thin line) is by May (1975).

4.4 Discrete time density-dependent models

We have looked primarily at the effects of density dependence in continuous time models. However, as we discussed earlier discrete time models are much more appropriate for insects that breed once per year. The time-delayed logistic model led to oscillations, but even more complex behavior is possible for the equivalent models in discrete time. At the time that this book is being written, the importance of this complex behavior in describing natural populations is an open question.

A class of models appropriate for describing animals (or plants) that live 1 year, reproduce, and then die, takes the form

$$N_{t+1} = F(N_t), \tag{4.33}$$

where the function F gives the population numbers next year in terms of this year's numbers. This equation has an implicit time delay of 1 year, so we expect behavior at least as complex as

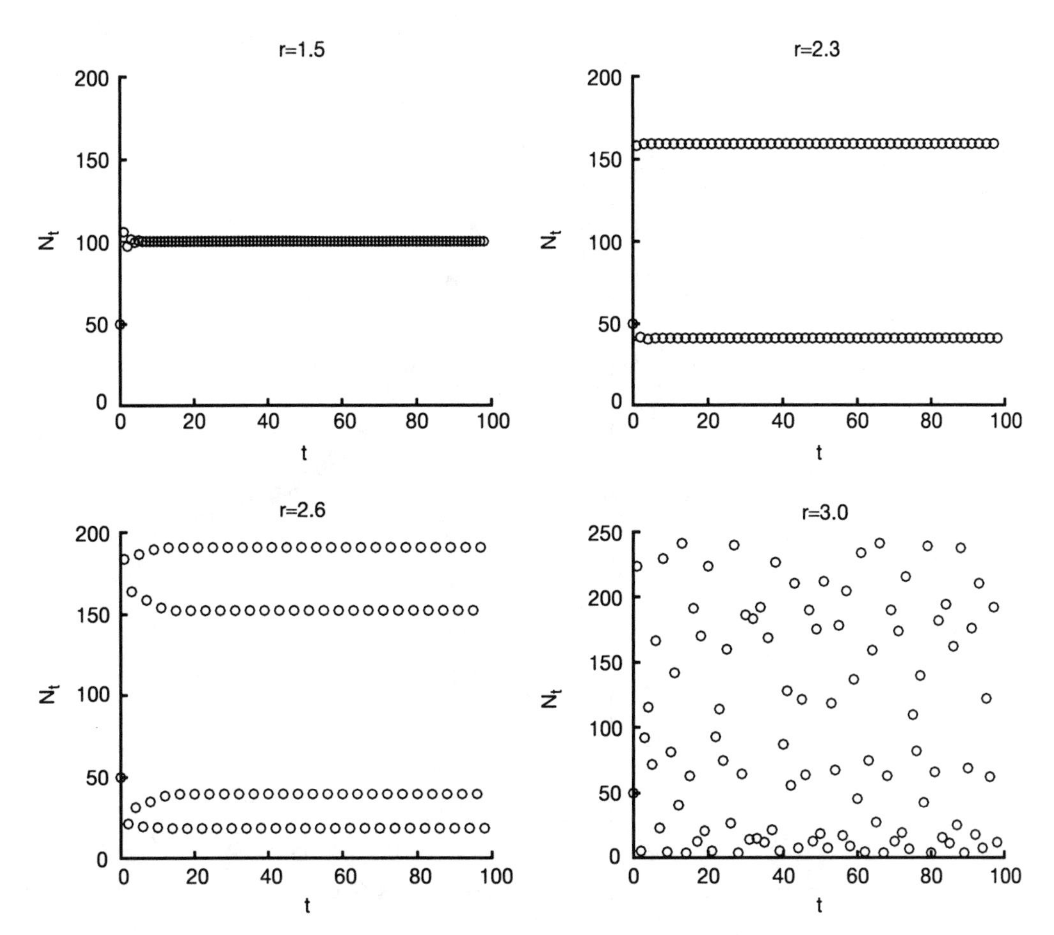

FIGURE 4.10. Dynamics of the Ricker model, which is given by equation 4.35, as a function of the parameter r. Each panel plots numbers against time, as found by solving the equation numerically. In each case the initial condition $N(0) = 50$, and the parameter $k = 100$. For $r = 1.5$ the population quickly approaches an equilibrium. As r is increased, the population has a 2-cycle at $r = 2.3$, a 4-cycle at $r = 2.6$, and chaos at $r = 3.0$.

for the time-delayed logistic. We will use computer simulations to understand the potentially complex behavior.

A number of models (e.g., May, 1974) describing the dynamics of populations in discrete time have been proposed, all sharing some qualitative features. The plot of the population next year versus the population this year has a single maximum. At low population levels, next year's population increases as a function of the current population, but at very high population levels,

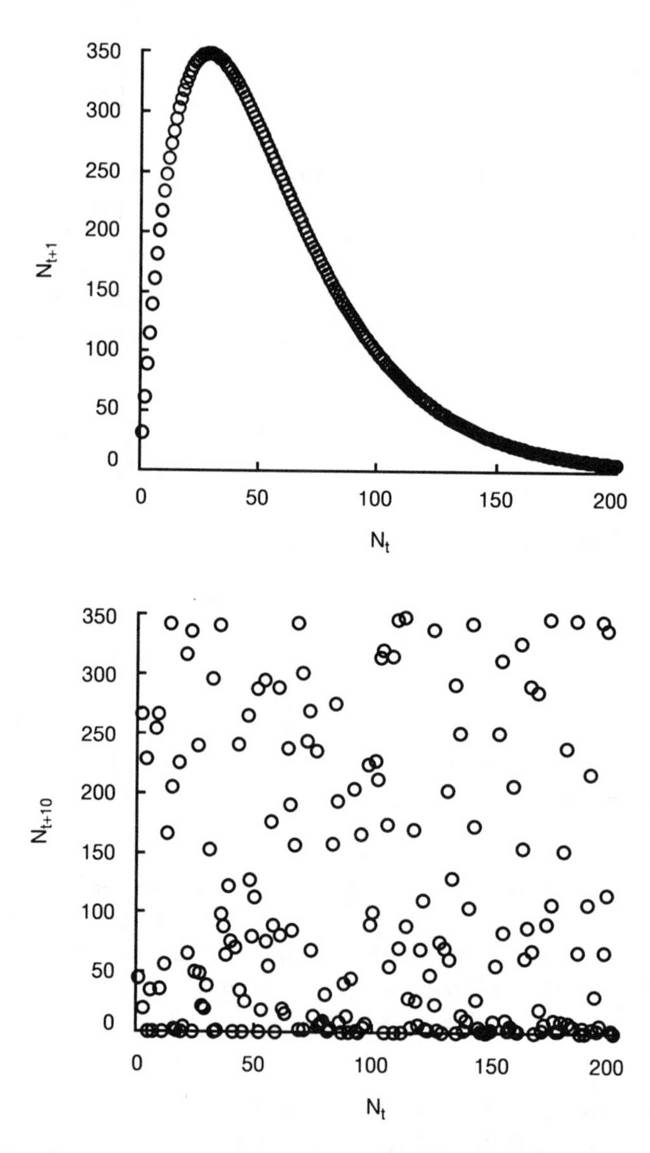

FIGURE 4.11. Dynamics of the Ricker model (equation 4.35) with $K = 100$ and $r = 3.5$. Population sizes after either 1 or 10 years are shown as a function of initial population sizes. In the top plot, one sees that it is easy to predict the population next year from the population this year. However, the bottom plot demonstrates that the population after 10 years is essentially impossible to predict from the population this year.

mechanisms of density dependence (intraspecific competition) reduce the population level next year. The mechanism of intraspecific competition may be through reduction of the food supply or through cannibalism. Three examples of models incorporating a description of density dependence are

Cannibalism is actually very common in nature.

$$F = N[1 + r(1 - N/K)], \tag{4.34}$$

$$F = N \exp[r(1 - N/K)], \tag{4.35}$$

and

$$F = \lambda N(1 + \alpha N)^{-\beta}. \tag{4.36}$$

All three of these models have behavior of astounding complexity. We illustrate some of the complex behavior possible for the Ricker (1954) model given by equation (4.35) in Figure 4.10. The stability analysis of the nontrivial equilibrium of this model is described in Box 4.3. For small values of r the model has damped oscillations. As r is increased there are sustained oscillations of period 2. As r is increased still further, there are cycles of period 4, then 8, then 16, and so on until a critical value of r is reached. For these higher values of r there is no simple cyclic behavior; the dynamics do not follow any simple pattern.

The behavior of the model for these higher values of r is called *chaotic*, and some of the aspects of chaos can be seen by looking at Figure 4.11. By plotting the population level next year versus the population level this year we emphasize that the dynamics are in fact deterministic and simple. However, if we plot the population level after 10 years versus the population level this year, as in Figure 4.11, a very different picture emerges. Even a small change in the population level now typically leads to a very large change in the population level after 10 years. Thus, although in this model we can predict the population level after 1 year, we cannot predict the population level after 10 years because we can never know the current population level well enough. This ability to predict over short time scales coupled with an inability to predict over long time scales is one of the central features of chaos.

Although chaos is not possible in a simple one-species continuous time model, it is possible in multispecies models with at least three species.

What would be the implications if chaos turned out to be a common feature of natural populations?

Box 4.3. Stability analysis of a single species discrete time ecological model.

We proceed with an example, the Ricker model,

$$N_{t+1} = N_t \exp[r(1 - N_t/K)].$$

- Find the equilibria by setting $N_{t+1} = N_t = N$, and solving for N. From the equation

$$N = N \exp[r(1 - N/K)],$$

we see that $N = 0$ or

$$1 = \exp[r(1 - N/K)],$$

which implies (using $\exp[0] = 1$) that

$$0 = [r(1 - N/K)].$$

We focus attention on the nontrivial solution of equation (4.3), the equilibrium $N = K$.

- Determine the stability of the equilibria using the procedure from Box 3.3. Linearize the model by computing dF/dN, where F is given by equation (4.35), and evaluating the derivative at the equilibrium $N = K$. The derivative is

$$\begin{aligned}\frac{dF}{dN} &= \frac{d(N \exp[r(1 - N/K)])}{dN} \\ &= \exp[r(1 - N/K)] + N(-r/K) \exp[r(1 - N/K)]).\end{aligned}$$

We then evaluate the derivative at the equilibrium:

$$\begin{aligned}\left.\frac{dF}{dN}\right|_{N=K} &= \exp[r(1 - N/N)] \\ &\quad + N(-r/N) \exp[r(1 - N/N)]) \\ &= 1 - r.\end{aligned}$$

The stability condition (see Box 3.3), that (a) be less than one in absolute value, is

$$|1 - r| < 1,$$

Box 4.3 (cont.)

which implies

$$0 < r < 2.$$

When $r > 2$, instability sets in because $\frac{dF}{dN}\Big|_{N=K}$ becomes less than -1. Instabilities arising because the 'growth rate' of deviations from equilibrium becomes less than -1 lead to oscillations. At each time step, the deviation from equilibrium will change sign, but become larger in absolute value. That oscillations result is confirmed by the numerical solutions in Figure 4.10.

What is not known is how important chaotic dynamics are in understanding the dynamics of natural populations. The techniques for studying this question are beyond the scope of this book. However, the laboratory population of Nicholson's blowflies illustrated in Figure 4.9 is one example of a population that some authors have suggested is chaotic. (Although chaos cannot be detected 'by eye', the irregularities that indicate the possibility of chaos are the varying heights of the population peaks and the varying times between peaks.)

4.5 Metapopulations

One could describe a population not by its total size, but by the fraction of available habitat sites it occupies, p. What we will do is describe the dynamics of this *metapopulation* consisting of a number of subpopulations. This kind of approach has been recently used extensively to understand metapopulations (Gilpin and Hanski, 1991; Hastings and Harrison, 1994). At any one site or location, the species colonizes the empty habitat, and then at some later time goes extinct. This cycle can then repeat, with the colonizers coming from other occupied sites. The transitions at one location are illustrated in Figure 4.12.

In the simplest description (Levins, 1969), the rate at which a population goes extinct from sites as the result of random pro-

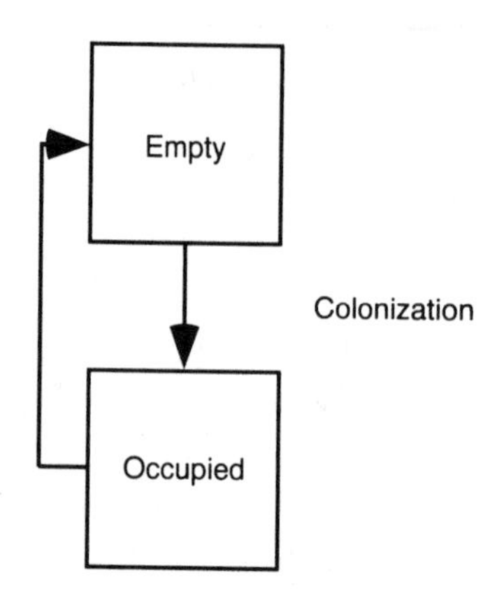

FIGURE 4.12. Transitions at one location in a simple model of a metapopulation.

cesses will just be proportional to the fraction of sites occupied ep, where e is a constant. (We choose the letter e to stand for extinction.) Similarly, the rate at which the population colonizes empty sites will be proportional to the product of the fraction of occupied sites times the fraction of empty sites, $mp(1-p)$, where m is a constant. (We ignore any effect on colonization arising from different locations of suitable habitat.) The total rate of change of the fraction of occupied sites will then be

$$\frac{dp}{dt} = mp(1-p) - ep. \tag{4.37}$$

The analysis of this model in the problems shows that a stable equilibrium arises (if colonization is large enough relative to extinction) at the metapopulation level, even though at the level of the subpopulation extinction is certain. This metapopulation concept has played a major role in conservation biology, even though its strict applicability has been questioned.

Conclusions

The consequences for dynamics do not depend on the *cause* of regulation, which could be cannibalism, limited food supply, or some other factor operating within a single species.

We have shown that the consequences of population regulation by factors operating within a single species can have very different consequences, depending on the time scale of regulation. If the regulation is instantaneous, the population reaches an equilibrium. Although we can numerically describe the population dynamics in this case, we have shown that quantitative conclusions from these models are suspect, although the qualitative conclusions are likely to be robust.

In contrast, if regulation is delayed, as with organisms that reproduce once per year and then die, or organisms where the delay until sexual maturity is significant, the dynamics can be much more complex, with cycles and chaos.

Unfortunately, the search for evidence of density dependence in natural populations has not yielded clear answers. Results from the survey by Hassell et al. (1989) are typical in this regard. Complications making the detection of density dependence difficult that we have not discussed in this chapter include the influence of stochastic forces, such as the weather, and the difficulty of obtaining large enough data sets. Perhaps the most important issue is elucidating the appropriate spatial and temporal scales for determining density dependence.

Problems

1. The model

$$\frac{dN}{dt} = rN\left[1 - (N/K)^{\theta}\right] \tag{4.38}$$

(where θ is a positive parameter that depends on the organism) has been proposed as an alternative to the logistic model (Gilpin and Ayala, 1973).

(a) Find the equilibria of the model, and determine their stability as described in Box 4.1.

(b) Graph the per capita growth rate

$$\frac{dN/dt}{N} = r\left[1 - (N/K)^{\theta}\right] \tag{4.39}$$

against N, for several different values of θ. (Use a value less than 1, a value greater than 1, and 1.) Discuss how changes in the value of θ alter the behavior of the model.

(c) Discuss briefly (one-half page) how you think this model may be superior to the logistic and to what taxa it might apply for different choices of θ. You may have to do additional reading to answer this question.

2. Here we analyze the metapopulation model, equation (4.37).
   (a) Find the equilibria of this model.
   (b) What is the condition required for the model to have an equilibrium with $p > 0$, and does this condition make biological sense?
   (c) Determine the stability of the equilibria in this model.
   (d) How is this model related to the logistic model?
3. Often, when a population level is too low, the population will decline, an effect known as the *Allee effect*. This effect can be incorporated in a simple model of the form

$$\frac{dN}{dt} = rN(N - a)[1 - (N/K)] \tag{4.40}$$

   where the positive parameter a is a threshold population level above which the population will grow.
   (a) Find the equilibria of the model.
   (b) Determine the stability of all the equilibria.
   (c) Graph the per capita growth rate in this model against the population size N.
   (d) Discuss the behavior of this model and contrast it with the behavior of the simple logistic model.
4. Variations in the simple logistic model have been used to study the effects of harvesting on the long-term dynamics of a species, for example, how many fish can be harvested without reducing the viability of the population. This situation can be described by a simple variation of the logistic model

$$\frac{dN}{dt} = rN[1 - (N/K)] - HN \tag{4.41}$$

   where H is the harvest rate.
   (a) Find the non-zero equilibrium of this model, which will depend on the value of H. What restriction on H is

necessary for this equilibrium to be positive? Discuss biologically why this condition makes sense.

(b) What happens to the population if H is larger than the value determined in part a?

5. In the text, we examined the behavior of the discrete time Ricker model. We claimed that the surprising and complex behavior of this model did not depend on the exact form of the model used. To show that the behavior is in fact robust, look at the behavior of a version of a discrete time logistic model

$$N_{t+1} = rN_t(1 - N_t/K) \tag{4.42}$$

for different values of r. The value of K does not in fact affect the form of the dynamics, so for simplicity you can use the value 1 for K. Thus, you should always have N between 0 and 1. The parameter K is not in fact the carrying capacity here.

(a) Construct a figure similar to Figure 4.10 using the values 1.5, 3.2, 3.5, and 3.9 for r.

(b) Construct a figure similar to Figure 4.11 using the value 3.9 for r.

Suggestions for further reading

The early work of Gause (1934, 1935) provides a large number of wonderful examples of population dynamics of laboratory organisms. Hutchinson's 1978 book provides an extensive historical discussion of the logistic equation and density dependence.

The role of nondimensionalization is discussed in Nisbet and Gurney's 1982 book. A further look at Nicholson's blowflies is in Gurney et al. (1980).

The question of chaos in ecology is reviewed by Hastings et al. (1993), which provides numerous further references. The important early paper by May (1976) on chaos in simple models is worth reading. One recent paper of note, by Costantino et al. (1995), shows the presence of two-cycles as in the simple models (although they in fact use a more complex model) in a laboratory population of flour beetles.

The early paper by Ricker (1954) on population regulation in fish is a classic in population ecology well worth reading. Crawley (1990) provides a detailed review of several studies of density-dependent population dynamics in plants. In the same symposium, Shepherd and Cushing (1990) review the more recent literature on population regulation in fish populations.

The metapopulation model discussed in the homework was originally introduced by Levins (1969). The metapopulation approach is extensively reviewed in both Gilpin and Hanski (1991) and Hastings and Harrison (1994).

5
Evolution of Life Histories

We have looked at the role of births and deaths at different ages in determining the dynamics of a population. The schedule of births, deaths, and emigration/immigration is known as the *life history* of a species. In addition to looking at the effect of the life history on the dynamics of a population, we can also ask how this life history arose. You are probably already familiar with plants and animals having wildly different life histories – some species are much longer lived and reproduce many more times than others.

Think of organisms with very different reproduction schedules.

In this chapter, rather than take a detailed look at the evolution of life histories, we will instead look at some particular questions. We will be illustrating approaches for studying the evolution of life histories rather than obtaining definitive answers.

5.1 Cole's result

One of the very first papers to discuss life histories from an evolutionary point of view was written by Cole (1954). He asked the question: why do plants reproduce when they do? The range of variation in nature is enormous, from annual plants that reproduce only once to plants such as oak trees which reproduce many times.

Cole's basic method was to ask under what circumstances would the growth rate of an annual and a perennial be the same. We thus begin by writing down equations for the numbers of an annual species or a perennial species after 1 year of population growth. Let B_A and B_P be the number of seeds (which in fact germinate the following year) produced per plant by the annual and perennial species, respectively. Similarly, let N_A and N_P be the numbers of annual and perennial plants, respectively. Let the survival rate of the perennial species be s. Then, for the annual species, we find

$$N_A(t + 1) = B_A N_A(t), \qquad (5.1)$$

and for the perennial species we find,

$$N_P(t + 1) = B_P N_P(t) + sN_P(t). \qquad (5.2)$$

We find the growth rates for the annual and the perennial species by finding $N_A(t + 1)/N_A(t)$ and $N_P(t + 1)/N_P(t)$, respectively.

What is the growth rate of each species? The growth rate of the perennial species is $B_P + s$, while the growth rate of the annual species is B_A.

Cole reasoned that the trade-off between being an annual and a perennial could be determined by equating the growth rates of plants following the two different strategies. His assumption was that the energy a plant puts into survival, i.e., into shoots or tubers or other vegetative structures, could instead be put into producing more seeds. The two growth rates are equal if

$$B_P + s = B_A. \qquad (5.3)$$

What is the largest s can be? Since s can be at most 1, the growth rate of the annual will definitely be larger than the growth rate of the perennial if the annual produces at least one more seed than the perennial, on average. Cole thought producing one more seed required far less energy than the structures necessary to survive from year to year, so he asked, why aren't all plants annuals? See if you can anticipate some of the arguments that have been advanced to counter Cole's reasoning before you read on.

What are the implicit assumptions in Cole's model that we have not discussed?

5.2 Extensions of Cole's model

One problem with the model we have just presented is that we have not separated out mortality of adults and juveniles. Although plants may produce many seeds, typically the survival rate of each seed is very low. Because the survival rate is so low, we expect that including mortality at the seed or seedling stage may have a large effect. Charnov and Schaffer (1973) extended the model we have just discussed by explicitly including mortality. We let the survival rate of juveniles for both types be s_j. The equations describing the dynamics of each type are modified such that the dynamics of an annual are then given by

$$N_A(t+1) = s_j B_A N_A(t) \tag{5.4}$$

and those of a perennial by

$$N_P(t+1) = s_j B_P N_P(t) + s N_P(t). \tag{5.5}$$

Now, what is the growth rate of each species? The growth rate of the perennial species is $s_j B_P + s$, while the growth rate of the annual species is $s_j B_A$. In this case the growth rates are equal if

$$s_j B_P + s = s_j B_A, \tag{5.6}$$

which we rearrange as

$$B_A = B_P + \frac{s}{s_j}. \tag{5.7}$$

What is the ratio s/s_j likely to be? To pick an extreme case, what is the typical germination probability and survival rate of a tree seed and seedling relative to the survival of a mature tree? The answer is that the ratio s/s_j may be very large, so for the annual to have the same growth rate as the perennial, B_A may need to be much larger than B_P. We therefore conclude that there may be some cases in which perennials would certainly be favored, and in fact, the problem may be to explain the circumstances under which an annual would be favored.

Keep thinking of difficulties with the simple model we have just developed.

5.3 Big bang versus iteroparous reproduction

The case we have just looked at is really a special case of comparing a *semelparous* species, one that reproduces just once, to an *iteroparous* species, one that reproduces many times. Semelparous, or 'big bang', reproduction is one of the more spectacular ecological strategies, with some plant species flowering once after a lifespan approaching a century. Examples of semelparous species include *Agave* sp. (century plants), some palm species, some bamboos, foxgloves, the fantastic Haeakala silversword, and salmon. Given our conclusion, based on comparing annuals and perennials, that it may be difficult to explain why an annual life history might be favored, we need to explain why big bang reproduction would evolve.

We will use another approach to look at this question based on the concept of reproductive value, following the work of Schaffer (1974). We introduced the notion of reproductive value earlier, and this will form the basis of our current investigation. We will base our approach on the following clever argument from optimization theory. We start at the end of a plant's life and work backward. As a function of the age and size of the plant, we determine the optimal reproductive strategy, and then work toward younger ages. The advantage of this approach is that the reproductive strategy at an older age cannot affect the behavior at a younger age.

Our entire development here assumes that natural selection has optimized the growth rate of the population. How does this fit with our discussion of population genetics?

How can an individual of age i contribute to future reproduction? There are two ways: births at age i and births at later ages. Note that we also have to discount births at later ages relative to the growth rate of the population. The value of current births is given just by the birth rate, b_i. We express the value of future births in terms of the reproductive value. Thus, the value of future births is given by the probability of surviving to the next age, P_i, times the reproductive value of an individual 1 year older relative to the reproductive value of a newborn. In symbols, the fitness is maximized at every age if

The 'currency' we use is the reproductive value of a newborn.

$$b_i + P_i \frac{V_{i+1}}{V_0} \tag{5.8}$$

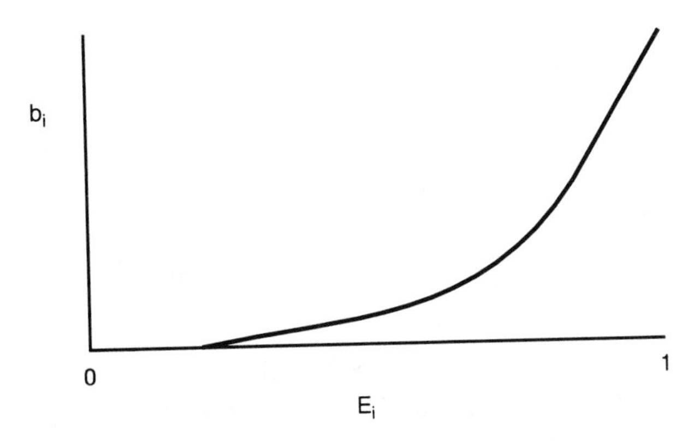

FIGURE 5.1. Schematic illustration of dependence of births at age i, b_i on reproductive effort E_i.

is maximized. This equation shows the virtue of our strategy of working backward – we can study the problem of evolution of a life history over all ages by optimizing a single age at a time. We assume, when trying to understand trade-offs at age i, that behavior at ages older than i is known, so the unknowns in equation (5.8) are the quantities b_i and P_i, the behaviors at age i.

We assume that there is a *trade-off*, so that increases in b_i lead to decreases in P_i and vice versa. We express the trade-off by using the concept of *reproductive effort* at age i, E_i, which is defined as the proportion of available energy resources allocated to breeding. We therefore assert that b_i should be an increasing function of E_i, as indicated in Figure 5.1.

Why do we assume that b_i is 0 for all values of E_i smaller than a critical value?

Similarly, we denote the contribution of future reproduction (conditioned on survival to the next age, of course) at age i by PV_i, so

$$PV_i = P_i \frac{V_{i+1}}{V_0}. \tag{5.9}$$

We can thus display the dependence of PV_i on reproductive effort in a similar graphic fashion.

Our goal is to maximize, at each age, the quantity given in (5.8), which is the sum of the two quantities we have just defined. We will first illustrate this in the case where the dependence of birth rate and future reproduction on reproductive effort are both convex. This case is illustrated in Figure 5.2. It is quite easy to see

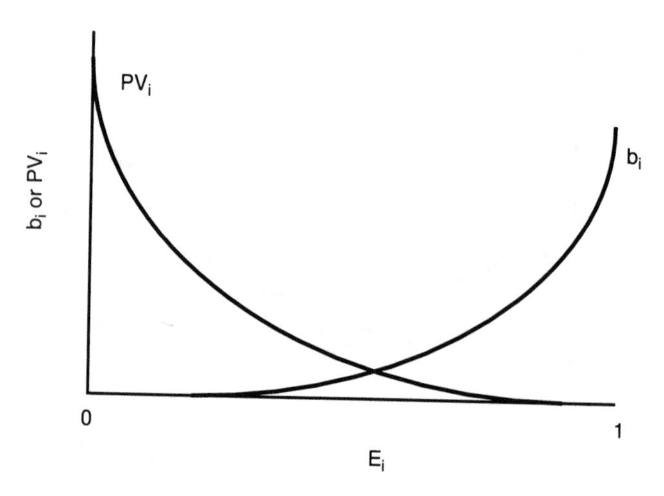

FIGURE 5.2. Schematic illustration of the dependence of births at age i (b_i) and future reproduction (PV_i) on reproductive effort E_i. Here, because both curves are convex, the maximum of their sum is at E_i equal to 0 or 1.

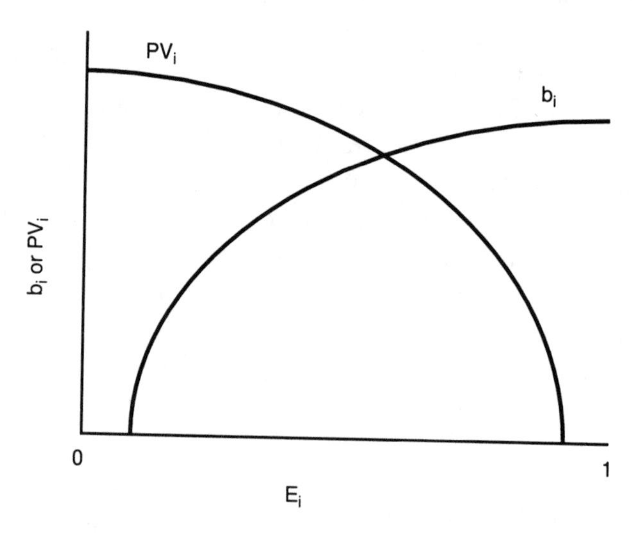

FIGURE 5.3. Schematic illustration of dependence of births at age i (b_i) and future reproduction (PV_i) on reproductive effort E_i. Here, because both curves are concave, the maximum of their sum is at a value of E_i between 0 and 1.

To find where the maximum occurs, try to see what happens to (5.8) near E_i equal to 0 or 1.

in this case that the maximum of (5.8) occurs either at $E_i = 0$ or $E_i = 1$. Thus, in this case semelparity is favored.

If, on the other hand, both of the curves are concave, as in Figure 5.3, we obtain a very different result. In this case (5.8) is

TABLE 5.1. Intensity of competition for pollinators in *Yucca* and *Agave* populations, classified by reproductive strategy (data from Schaffer and Gadgil, 1974). The way competition was measured is described in the text. It is clear that competition is much more intense in the semelparous species than in the iteroparous species. SD, standard deviation.

Semelparous (big bang)			Iteroparous		
Species	*N*	Competition	Species	*N*	Competition
Agave schottii	95	0.22	*Agave parviflora*	193	-0.01
A. utahensis	86	0.24	*Yucca standleyi*	100	0.00
A. deserti	84	0.15	*Y. utahensis*	162	0.01
A. chrysantha	20	0.18	*Y. glauca*	117	-0.01
A. palmeri	48	0.32	*Y. elata*	84	0.01
Yucca whipplei	23	0.08			
Mean ± SD		0.21 ± 0.09	Mean ± SD		0.00 ± 0.01

clearly maximized at some value of E_i between zero and one. In this case iteroparity is favored.

In Figure 5.3, compare the value of (5.8) at the point where the curves cross to the value of (5.8) near either endpoint.

This theory was used by Schaffer (1974) to try to understand the distribution of iteroparity and semelparity among 11 species of yuccas and agaves. He reasoned that if plants compete for pollinators, the curve describing the dependence of birth rate on reproductive effort would be convex, as in Figure 5.1, because the plants producing the tallest flowering stalks would be much more successful in attracting pollinators. As a measure of how important intraspecific competition for pollinators was – whether taller plants were more successful at getting pollinated – Schaffer (1974) computed the regression of the percent of flowers fertilized against the quantity $(H - \bar{H})/\bar{H}$, where H is the height of the flowering stalk of an individual plant and $\bar{H}$ is the mean height of the flowering stalks of other plants of the same species.

What result would confirm the theory?

The 11 species examined included 6 that were semelparous and 5 that were iteroparous. As listed in Table 5.1 for the 6 semelparous species the slope of the regression defined in the previous paragraph had a mean of 0.21 ± 0.09. In contrast, for the 5 iteroparous species the regression had a mean of 0.0 ± 0.01. Thus, competition for pollinators is apparently much more intense in the semelparous species, in accordance with the theory presented here.

5.4 Evolution of dispersal

Another important life history characteristic is dispersal. There have been many approaches to determining an optimal level of dispersal, but here we give just one example, from Hamilton and May (1977), which illustrates the use of ESS arguments as well.

The concept of an ESS (evolutionarily stable strategy) was defined in the previous chapter.

Consider a species inhabiting an environment consisting of a fixed number of sites at which only a single adult can survive. At the end of the season a fixed number of offspring, m, are produced. Of these, a fraction v are programmed to disperse, while a fraction $1-v$ remain at the home site. Assume that the adult always dies, and that only a fraction p of the migrants reach a new site. From those individuals at a given site at the start of the season, one individual is chosen (at random) to breed successfully.

What is the ESS? First, we show that a complete absence of dispersal is clearly not an ESS. So long as an individual keeps at least one offspring at its home site, it can successfully invade a population with no dispersal. If none of the other organisms disperses, the single individual remaining at its natal site will always survive. No matter how small the chance of success of the disperser, eventually sites will be taken over by the dispersing type, which can never be displaced from a site it occupies. This simple argument shows that the ESS must incorporate some level of dispersal.

Using the approach we outlined earlier for finding a mixed ESS in a system with two discrete choices, we can find the ESS here. It is difficult to find the exact dispersal level that is the ESS, so we merely present the answer here. It is

$$v* = \frac{1}{2-p} \tag{5.10}$$

if the number of sites and propagules per parent is large. Thus dispersal pays, even in a stable habitat.

It is easy to show that dispersal is advantageous in a habitat that is not constant.

This result is important and interesting because adaptations promoting dispersal are ubiquitous within the plant and animal kingdoms. Seeds are blown about by the wind or carried attached to the fur of animals, or fruits are eaten and the seeds deposited elsewhere in the feces of the animal. Larvae of marine animals are carried long distances by ocean currents. Adult insects often can fly great distances or even be carried by wind currents. It is also not surprising that such a universal adaptation can be explained in several different ways.

Conclusions

The examples we have discussed show that understanding the natural world requires the use of evolutionary in addition to ecological ideas. Every ecological interaction has been shaped by the forces of natural selection. The outcome of natural selection cannot be described by simple optimization, but through the use of the ESS concept, many of the adaptations of organisms can be understood.

Problems

1. What effect would environmental variability have on dispersal? Answer this for the following simple situation. Assume that at any site occupied by an annual plant, there is a 10% chance that in any given year all the plants die before they are able to reproduce. How would a phenotype fare that does not disperse at all compared to one which disperses? Use verbal arguments.

2. Another strategy that has evolved in plants to cope with environmental variability is dormancy: some seeds do not germinate the year after they are produced, but remain in the soil (or seed bank) until a later year. Explain how this strategy is a complement to dispersal for coping with environmental variability.

3. In light of the first two problems, discuss how the addition of variability and dormancy might change Cole's original argument about the evolution of perennials and annuals.

Suggestions for further reading

Two different books with the same title, *The Evolution of Life Histories*, both appeared in 1992 – one by Stearns, the other by Roff. Both are extensive syntheses. Cole's original 1954 paper is a classic well worth reading.

Models and data pertaining to the evolution of dispersal are reviewed in Johnson and Gaines (1990).

Part II

Interacting Species

6

Interactions Between Species

In this chapter we begin our examination of interactions between species. For two reasons, we will focus only on the interactions between two species.

- There are some natural systems that come close to being two species systems, such as the predator–prey interaction between wolves and large ungulates or the interaction between hosts and parasitoids.
- Before we can hope to understand systems with more than two species, we will need to understand systems with two species. Two-species models are simpler both biologically and mathematically.

We start by outlining the different kinds of interactions possible between pairs of species. If both species use a common resource, or inhibit each other, each will have a negative effect on the other. We call this kind of interaction *competition*.

A second kind of interaction involves the consumption of one species by another. We will call this a *predator–prey* system, where the predator eats the prey. The interaction is predation. A similar situation would be one where an animal eats a plant, so more generally we can speak of *exploiter–victim* interactions.

A third kind of interaction, which we do not discuss in detail here, is one in which each species benefits the other. One example is the relationship between a plant species and a pollinator whereby the plant benefits by having its seeds pollinated and the pollinator benefits by collecting nectar (a food source). This is called *mutualism*. These three classes do not encompass all possible interactions, but are a useful categorization.

How would you classify the interaction between humans and a potentially fatal disease?

Where do decomposers fit into this classification?

The paper by Hairston, Smith, and Slobodkin (1960) has had such an impact that it is often referred to by their initials, HSS.

As already noted, a long-standing question in ecology is what actually regulates populations in nature. This question was expressed quite forcefully by Hairston, Smith, and Slobodkin (1960), who suggested that since the earth is 'green' – live plant material is abundant – then predation, not food limitation or competition, must act to regulate the populations of herbivores (consumers of plant material). This appealing argument may be correct, or it may be that much of the green and apparently edible plant material is in fact not easily consumed because it is protected by either physical or chemical means. HSS goes on to state that the consumers of herbivores are themselves limited by their food supply; they are regulated by competition. The ultimate goal of our study of population interactions is to shed light on this question of what regulates populations: to examine the consequences of different kinds of regulation and elucidate ways to determine which form of regulation is occurring.

In this chapter, we set up the overall approach we will use to examine population interactions that could serve to regulate populations in nature. We will consider both the intuitive concept of stability and a precise definition of stability. We also look at specific steps we will use in analysis and computation of the stability of equilibria.

6.1 Two-species models

Before turning to models of the specific interactions we have defined, we explore some general features of two-species models, as well as some issues central to our attempts to understand the interactions between two species. The general form of the models we will examine in succeeding chapters is the natural extension of the one-species models we have examined earlier. Let N_i be the number of individuals in species i. In a continuous time model, we will assume that

How would the different kinds of interactions between species be reflected in the per capita growth rates f_i?

$$\frac{dN_1}{dt} = N_1 f_1(N_1, N_2) \tag{6.1}$$

$$\frac{dN_2}{dt} = N_2 f_2(N_1, N_2), \tag{6.2}$$

where f_i is the per capita growth rate of species i as a function of the numbers of species 1 and 2. This framework can be used to examine the different interactions of competition, predation, mutualism, and even diseases. We will use an analogous discrete time model later. Before we can focus on the consequences and causes of competition, predation, and diseases, we need to focus on the tools we will use to explore these models.

Concept of stability

Underlying all our investigations of two-species systems is the concept of *stability*. We have used this concept in our one-species models, and will define it precisely in the next section. One rationale underlying the use of stability is the claim that the systems we see in nature correspond to stable solutions of the models we use to describe these systems. This claim is controversial for both biological and mathematical reasons.

The biological controversy over stability is essentially the question of how stable natural systems really are. Connell and Sousa (1983) attempted to provide an empirical answer to this question. Their data, given in Figure 6.1, show that variability in natural populations often is quite high. What has remained a subject of controversy is the cause of the variability that is observed – is it the result of ecological processes at the level of the population, or is it a reflection of environmental variability?

The mathematical controversy will become clearer when we give our definition of stability, but the problem is that our notion of stability, as we have seen earlier, strictly relates only to behavior near equilibrium. Our ultimate goal is to understand the dynamics of two-species systems, and not just their behavior near equilibrium. However, there still is a rationale for beginning with stability

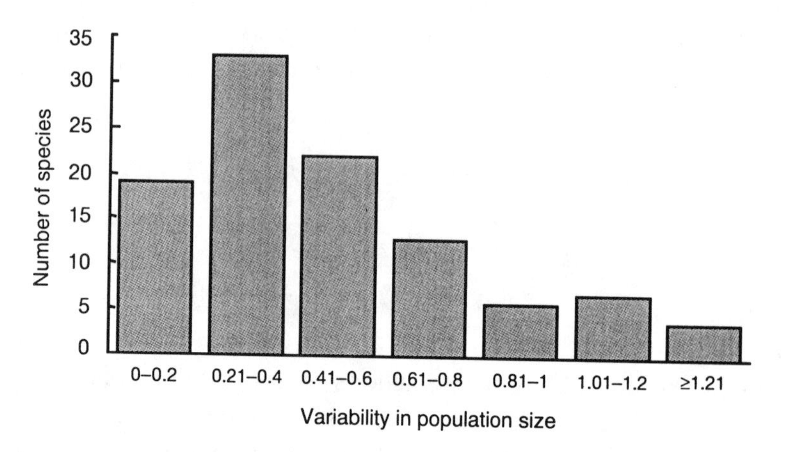

FIGURE 6.1. Population size variability (from Connell and Sousa 1983). This is a plot of the number of species exhibiting different levels of population stability, defined as the standard deviation of the logarithm of population sizes for the species. The original data on population sizes were obtained from a literature survey.

analysis. The first step toward understanding the dynamics of a system is to understand its possible equilibrium behavior, because complete solutions are almost always impossible. In contrast to one-species models, there are virtually no two-species models for which we can write down an explicit solution.

6.2 Definition of stability

We will first precisely define stability in the context of equilibrium behavior. The intuitive concept is that an equilibrium is *stable* if the system returns to the equilibrium when perturbed. To make this definition more precise we need to define carefully the notion of return to equilibrium and perturbation. A system is *locally stable* if for some sizes of perturbation – which might be arbitrarily small – away from the equilibrium, the system stays near the equilibrium. If in addition, the system approaches the equilibrium through time, it is said to be *locally asymptotically stable*. This notion is biologically suspect because we do not expect arbitrarily small perturbations, but the concept is mathematically convenient. When we use this notion in our biological reasoning we are as-

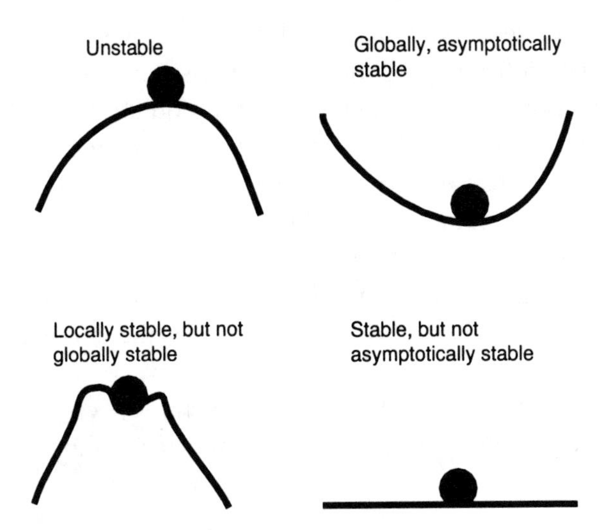

FIGURE 6.2. Concepts of stability used in analyzing ecological models. The ball 'rolls downhill'.

suming that the system can, in fact, withstand perturbations that are not arbitrarily small.

We can also define *global stability*, where we allow perturbations of arbitrary size within the confines of a particular model. The notion of global stability can be used when we do not want to restrict attention to arbitrarily small perturbations, but we must remember that we are still working within the confines of a particular model. Examples of biological circumstances that would cause a system to leave the confines of a particular model could be the addition of a species not included in the model, or the complete removal of a species.

Large perturbations might be fire, or a hurricane.

We use perturbation for temporary effects. Long term changes, like global climate change, are systematic changes in a model.

6.3 Community matrix approach

We will develop the general approach used to determine local asymptotic stability, which depends on the mathematical tools we introduced to study age-structured models. Because we are focusing on local asymptotic stability, we need only consider arbitrarily small perturbations. Thus, we only need to consider the dynamics of the model near equilibrium, as we did for the one-species model.

Recall that we linearize (look only at linear terms) because at equilibrium, $n_i = 0$, and near equilibrium $n_i \ll n_i^2$. Since linearizing only works near equilibrium, this is where our assumption of small perturbations is used.

We approximate equation (6.2) by linearizing near equilibrium. The deviation from equilibrium for each species is $n_i = N_i - \hat{N}_i$. We approximate the dynamics near the equilibrium by a linear model. We then use the theory of eigenvectors and eigenvalues developed in Chapter 2 to look at age-structured growth, to compute the growth rates of perturbations away from equilibrium. If all perturbations shrink to zero, the equilibrium is stable. If any perturbation grows, the equilibrium is unstable. The matrix that we use in linearizing the model (Box 6.1) is the *community matrix*, the matrix describing the effects of species on their own growth rate and the growth rate of interacting species. The entire procedure for determining stability is summarized in Box 6.1.

6.4 Qualitative behavior of the community matrix

We can gain some insight into the dynamics of two-species systems by considering the qualitative nature of the community matrix in two-species models. First, let us determine the signs of the entries, when they can be determined, in the community matrix for two species. In a competition model we assume that all the entries are negative. In a predator–prey model, we would assume that the effect of the prey on the predator is positive, while the effect of the predator on the prey is negative.

This information is useful because of the following facts, which can lead to some simple stability conditions:

These facts about the relationship between the trace and determinant and eigenvalues of a matrix hold for larger matrices, except the determinant of an $n \times n$ matrix equals $(-1)^n$ times the product of the eigenvalues. These relationships can help to understand qualitative features of stability of more complex systems.

- The *trace* of a matrix (the sum of its diagonal elements) is equal to the sum of its eigenvalues.
- The determinant of a matrix is equal to the product of its eigenvalues.

Thus, we see that if both eigenvalues are negative, the determinant is positive and the trace is negative. If one eigenvalue is positive and one is negative, the determinant is negative. If both eigenvalues are positive, the trace must be positive. From this we can determine simple stability conditions, as summarized in Box 6.2.

Box 6.1. Computing equilibria and stability in a two-species model.

Begin with a model of the form

$$\frac{dN_1}{dt} = N_1 f_1(N_1, N_2) \equiv F_1(N_1, N_2)$$
$$\frac{dN_2}{dt} = N_2 f_2(N_1, N_2) \equiv F_2(N_1, N_2),$$

so we have defined both per capita growth rates f_i and total growth rates F_i.

- Determine equilibria by setting F_i to be 0 and solving for $\hat{N}_1$ and $\hat{N}_2$. (Note that if $F_i = 0$, then either $N_i = 0$ or $f_i = 0$.) You must solve both equilibrium equations simultaneously, which may not be easy and in fact may be impossible. There will typically be more than one equilibrium; one will be the trivial one with both species absent ($\hat{N}_1 = \hat{N}_2 = 0$). There can be equilibria with one species absent ($\hat{N}_1 = 0, f_2(0, \hat{N}_2) = 0$ or $\hat{N}_2 = 0, f_1(\hat{N}_1, 0) = 0$). There may also be nontrivial equilibria with both species present, so $f_1(\hat{N}_1, \hat{N}_2) = 0$ and $f_2(\hat{N}_1, \hat{N}_2) = 0$.
- Linearize the model about each equilibrium. Describe the rate of change of the deviation from equilibrium, $n_i = N_i - \hat{N}_i$, near equilibrium by the equation

$$\begin{pmatrix} \frac{dn_1}{dt} \\ \frac{dn_2}{dt} \end{pmatrix} = \begin{pmatrix} \alpha_{11} & \alpha_{12} \\ \alpha_{21} & \alpha_{22} \end{pmatrix} \begin{pmatrix} n_1 \\ n_2 \end{pmatrix},$$

which in matrix form is

$$\frac{d\vec{n}}{dt} = A\vec{n},$$

where $\vec{n}$ is the vector of population sizes.

- Compute the elements of the community matrix A as for one species, using the formula

$$\alpha_{ij} = \frac{\partial F_i}{\partial N_j},$$

Box 6.1(cont.)

where the partial derivatives are evaluated at the equilibrium for which we are trying to determine stability.

- Determine stability by computing the eigenvalues λ of the matrix A as the solutions of the equation

$$(\alpha_{11} - \lambda)(\alpha_{22} - \lambda) - \alpha_{12}\alpha_{21} = 0.$$

- If the real part of both solutions λ is negative, the equilibrium is stable. If the real part of either solution λ is positive, the equilibrium is unstable.

(We will discuss complex numbers when we discuss predator–prey models. This will explain our use of the term 'real part'.)

Box 6.2. Stability conditions for a 2×2 matrix.

Both eigenvalues of the 2×2 matrix

$$\begin{pmatrix} \alpha_{11} & \alpha_{12} \\ \alpha_{21} & \alpha_{22} \end{pmatrix}$$

have a negative real part if and only if

$$\alpha_{11} + \alpha_{22} < 0$$

and

$$\alpha_{11}\alpha_{22} - \alpha_{12}\alpha_{21} > 0.$$

If either of the inequalities is violated, then the equilibrium is unstable.

Summary

Our goal is to understand what regulates populations. If regulation results from interactions with other species, the process could be competition or predation. The concept of stability will play a central role in our investigation of interactions between species. We have presented both a way to compute stability for two-species models and a discussion of the rationale for using stability as a central theme. It is important to recognize the limitations of using stability as the means to understand the dynamics of species in nature – there are both biological and mathematical problems. In the following chapters, we will use approaches based on stability and additional ways of understanding the dynamics of interacting species.

Problems

1. Discuss why concepts of stability of mathematical models are likely (or are not likely) to be useful approaches when trying to understand why we observe the communities we see in nature. If you see difficulties with the concept of stability, can you suggest alternative concepts?
2. In this chapter we emphasize the role of two-species interactions for studying the dynamics of interacting species. Explain both the advantages and potential pitfalls of focusing on the interactions between two species, as opposed to looking at more than two species. (You need to indicate which species is 'species 1' and which is 'species 2'.)
3. Later, when we actually analyze models of interacting species, the following will be important: classify models as predator-prey, competition, or mutualist by determining the signs of $\frac{\partial f_1}{\partial N_2}$ and $\frac{\partial f_2}{\partial N_1}$, the change in per capita growth, as the numbers of the other species are changed.

Suggestions for further reading

The classic paper by Hairston, Smith and Slobodkin (1960) sets the stage for any discussion of population regulation. For more recent work on this subject see the edited books by Carpenter and Kitchell (1993) and Polis and Winemiller (1996).

Concepts of stability in an ecological context are discussed in Lewontin (1969) and Holling (1973).

The survey paper of Connell and Sousa (1983) provides important data on stability, with the paper of Hanski (1990) giving an update on this important topic. Issues related to chaos and stability are discussed in Hastings et al. (1993).

The approach to modeling two-species interactions presented here began with the classic work of Lotka (1926, 1932) and Volterra (1926, 1931). May (1975) presents a more modern, in-depth treatment of this topic.

7
Competition

What happens when two species are grown together? One of the first careful experiments of this kind was performed by Gause (1935). He grew two different species of *Paramecium* both separately and together under laboratory conditions. The results he obtained are graphed in Figure 7.1. As you might guess, the number of each species was lower when the two species were grown together than when each was grown individually. Similar experiments have been performed many times since.

We would like to understand this process of competition. Can we predict when the outcome will be the one obtained by Gause (1935) in which both species coexist? What would cause one species to eliminate the other instead? Can we predict, or understand, how much the equilibrium level of each species would be reduced by the other? In the experiment by Gause, the population of each species approached its equilibrium value more or less monotonically. Is this the expected result? To answer these and other questions about the dynamics of competition, we turn to a modeling approach.

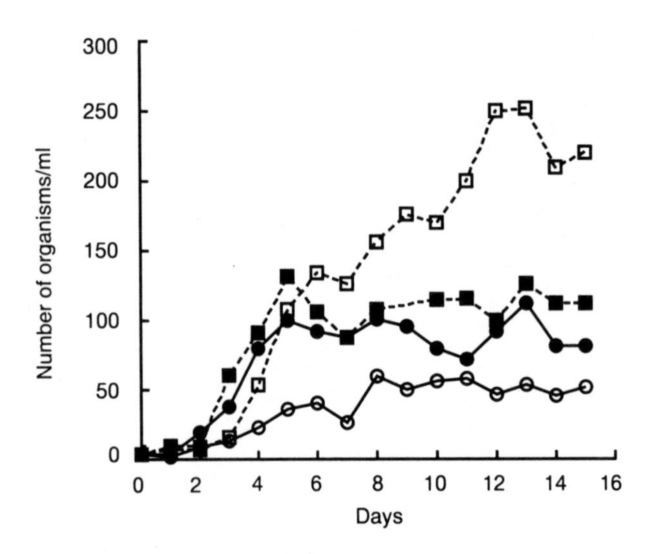

FIGURE 7.1. Competition between two laboratory species of *Paramecium*. The numbers of *P. bursaria* when grown alone are indicated by open boxes, and the numbers of *P. caudatum* when grown alone are indicated by closed boxes. The results of growing the two species together are also given, with the number of *P. bursaria* indicated by open circles, and the number of *P. caudatum* by closed circles. The numbers of both species are lower when they are grown together, and they reach an approximate equilibrium. The experimental protocol included removal of a fixed fraction of the individuals each day (data are from Gause 1935).

7.1 Lotka–Volterra models

Gause's work on competition was motivated by the Lotka–Volterra competition models, which can be viewed as the natural extension of the logistic model to two species. The logistic model can be written as

When initially presented by Lotka and Volterra, this model was justified phenomenologically, meaning that no underlying biological meaning was attached to the coefficients a_{ij}. There was no prescription for measuring these competition coefficients other than fitting the model to data. Since then, there have been some attempts to derive these competition coefficients from other models.

$$\frac{dN}{dt} = rN(1 - \frac{N}{K}) \tag{7.1}$$

$$= rN(1 - a_1 N), \tag{7.2}$$

where $a_1 = 1/K$.

We now look at a system with two species, where we let N_i be the number of individuals in species i, r_i is the intrinsic growth rate of species i, and include four positive constants, a_{ij} which represent interspecific and intraspecific density dependence. Here we start with equation (7.2). For each species we replace the term a_1 by a_{ii} to represent the effect of each species on itself, and add a second term to each equation with the coefficient a_{ij} representing the effect of species j on species i. Extending the logistic model, we obtain

$$\frac{dN_1}{dt} = r_1N_1(1 - a_{11}N_1 - a_{12}N_2) \qquad (7.3)$$

$$\frac{dN_2}{dt} = r_2N_2(1 - a_{21}N_1 - a_{22}N_2). \qquad (7.4)$$

At times it is more useful to include explicitly the carrying capacity of each species, and write this pair of equations as

In this formulation, α_{12} gives the relative effect of species 2 on the population growth rate of species 1, as compared to the effect of species 1 on its own population growth rate.

$$\frac{dN_1}{dt} = \frac{r_1N_1}{K_1}(K_1 - N_1 - \alpha_{12}N_2) \qquad (7.5)$$

$$\frac{dN_2}{dt} = \frac{r_2N_2}{K_2}(K_2 - \alpha_{21}N_1 - N_2). \qquad (7.6)$$

We now look at the solutions of this model, trying to relate qualitatively different biological outcomes (different species surviving) to different assumptions about the species interactions (as reflected in the parameters of the model).

Graphical approach

One of the most powerful approaches for understanding the dynamics of two-species systems is the graphical approach. In this approach, we focus on a *phase plane* (Figure 7.2), where we draw the dynamics of species through time, but ignore the time axis, and look only at the species numbers.

The phase plane is the projection of a graph of the numbers of two species against time in three dimensions onto the two dimensions consisting of just the species numbers.

Why are we only interested in nonnegative values of N_1 and N_2?

The first step is to draw the isoclines (as they are known in ecology, although more properly called nullclines) for the model under consideration. Isoclines are the curves along which the rate of change of the population of a species is zero, i.e., $\frac{dN_i}{dt} = 0$ for $i = 1$ or 2. Thus, for example, the growth rate of species 1 is zero if

$$\frac{r_1N_1}{K_1}(K_1 - N_1 - \alpha_{12}N_2) = 0. \qquad (7.7)$$

This isocline equation has two solutions:

$$N_1 = 0 \qquad (7.8)$$

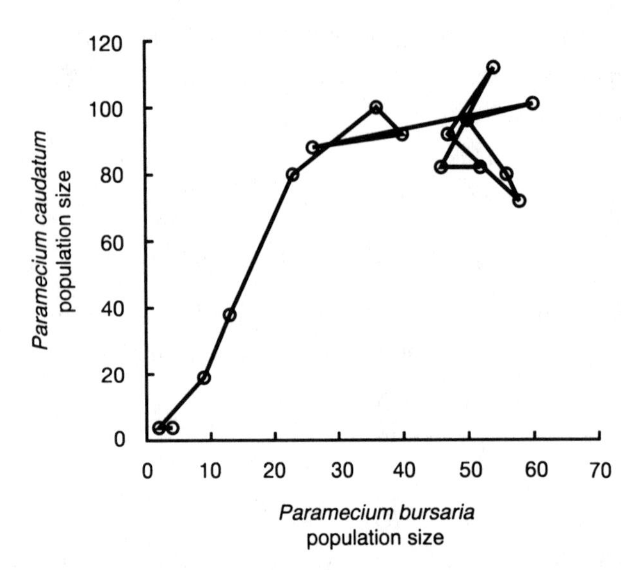

FIGURE 7.2. Phase plane depiction of competition between two laboratory species of *Paramecium*. The data from Figure 7.1 for the two species grown together are shown on a phase plane. The initial conditions are in the lower left, and the line joins circles which each represent numbers of the two species on a given day. As this is a phase plane, the time axis is implicit.

or

$$(K_1 - N_1 - \alpha_{12}N_2) = 0. \tag{7.9}$$

The first solution lies along the axis, so we will focus on the solution (7.9). To graph this equation, notice first that it is the equation of a straight line. A straight line is specified by two points. What two points that lie on this line are easy to find on a graph?

One point along (7.9) that is easy to find is $N_1 = 0, N_2 = K_1/\alpha_{12}$. What is the other?

We then graph the isocline for species two in a similar fashion, producing Figure 7.3. Equilibria occur at points where the isoclines for the two species cross, because these are the points where the growth rates of both species are zero. Equilibria also occur where an isocline for species 1 crosses the axis $N_2 = 0$, because the growth rate of species 2 is zero when $N_2 = 0$.

How many equilibria are there in Figure 7.3?

To proceed further we need an indication of how the species numbers change with time. Although time is not explicitly included in the phase plane as an axis, we indicate changes through time by vectors showing how the numbers in each species would change if the number of each species was given by the coordi-

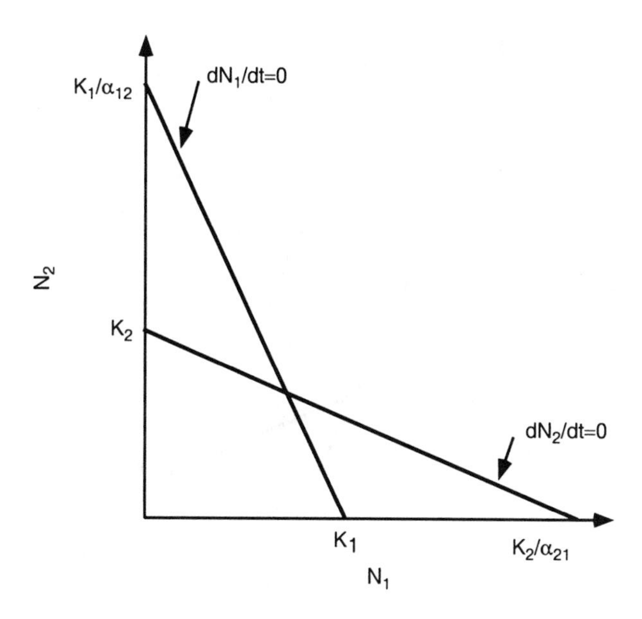

FIGURE 7.3. Isoclines in the phase plane of the Lotka–Volterra model for competition.

nates on the phase plane. From the equations, we see that increasing the numbers of either species reduces the per capita growth rate of both species. Thus, in the phase plane, in the upper-right portion the growth rate of both species is negative. We indicate this in Figure 7.4 both by labeling this portion and by drawing a vector showing that the change in numbers through time leads to a movement down and to the left.

We can then proceed in a similar fashion to label the other three portions of the phase plane to indicate the direction of change in species numbers. This is illustrated in Figure 7.5. In this case, we conclude from the phase plane that the species ultimately coexist, approaching the equilibrium at the center of the phase plane. Note that at this equilibrium each species is at a lower population level than it would be if it were by itself.

In drawing Figure 7.5, the outcome depended on the relative positioning of the carrying capacities, K_i, and the relative effects of the species on each other, α_{12}, α_{21}. There are four different cases possible, with three of them qualitatively different. By 'different cases', we refer to the relative positions of the isoclines. We illustrate two other cases in Figure 7.6 and Figure 7.7.

Figure 7.5 is just one possible placement of the isoclines. Try to think of others, leading to different outcomes of competition, before you go on.

There are two cases which are similar because one corresponds to the elimination of species 1 by species 2, while the other corresponds to the elimination of species 2 by species 1.

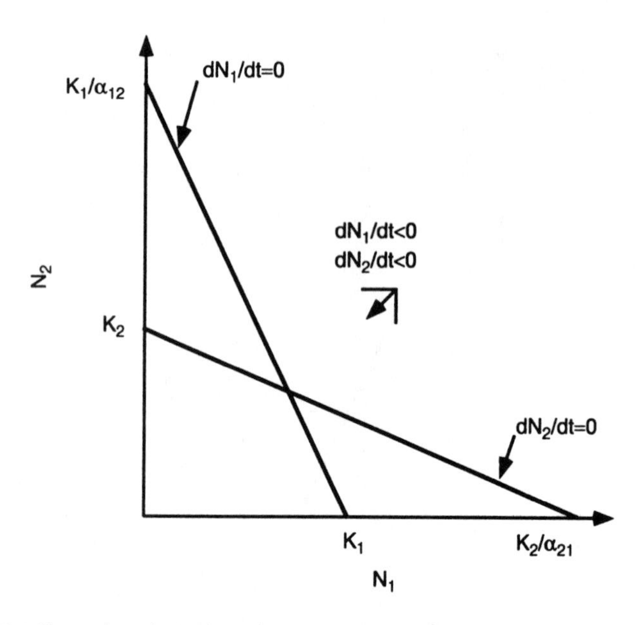

FIGURE 7.4. Phase plane for Lotka–Volterra competition. The next step in completing the phase plane in Figure 7.3 is drawing the direction of the change in species numbers. This is done for the upper right of the phase plane, where the numbers of both species are declining.

In Figure 7.6, from the directions of the arrows, we conclude that species 1 is eliminated and only species 2 remains. In Figure 7.7 we conclude that the stable outcome is the elimination of either species 1 or species 2, with the outcome depending on the initial conditions. The steps we have used in a general phase plane analysis are summarized in Box 7.1.

Another way to view the dynamics of competition is to solve the equations describing the dynamics on a computer and plot the numbers of each species through time. This is illustrated in Figures 7.8 through 7.10. The outcome in each case corresponds to what we would predict based on our phase plane analysis. Note that to solve the model by computer, we first have to pick specific parameter values.

The importance of our phase plane analysis, as opposed to just doing numerical solutions, is that the phase plane analysis can tell us the outcome of competition for all parameter values, while numerical solutions only tell us the outcome for each specific set of parameters for which we plot the solutions. Thus, even the pervasiveness of computers does not reduce the importance of the phase plane approach developed here.

From the phase portraits we conclude that the conditions for coexistence are those met in Figure 7.5, namely

$$K_1 < \frac{K_2}{\alpha_{21}} \tag{7.10}$$

Box 7.1. Steps in phase plane analysis.

To do a phase plane analysis of a model of the form

$$\frac{dN_1}{dt} = N_1 f_1(N_1, N_2) \qquad \frac{dN_2}{dt} = N_2 f_2(N_1, N_2)$$

we have employed the following steps.

- Determine the isoclines. The nontrivial isocline for species 1 is given by the equation $f_1(N_1, N_2) = 0$. The nontrivial isocline for species 2 is given by the equation $f_1(N_1, N_2) = 0$. Plot the solutions of these two equations on the N_1, N_2 plane, focusing only on nonnegative population sizes N_1 and N_2. Label each isocline as in Figure 7.7.
- Find the equilibria. The equilibria are the points where the isoclines for the two species cross, or where the isocline for species 1 intersects the line $N_2 = 0$, or where the isocline for species 2 intersects the line $N_1 = 0$.
- Find the signs of the rates of change of population size in the different parts of the phase plane. The isoclines will have divided the phase plane into different regions. Using both the model equations and the knowledge that the rate of change of each species is zero along its isocline, determine the signs of the population change $\frac{dN_1}{dt}$ and $\frac{dN_2}{dt}$ in each region of the phase plane. Write these on the phase plane.
- Draw arrows indicating the direction of population changes on the phase plane. If $\frac{dN_1}{dt}$ is positive, the arrow points to the right; if it is negative the arrow points to the left. If $\frac{dN_2}{dt}$ is positive, the arrow points up; if it is negative the arrow points down. Combine this information to determine the direction of population change. Draw the arrows indicating the direction on the phase plane.

From the phase plane, you can often infer the fate of interacting populations.

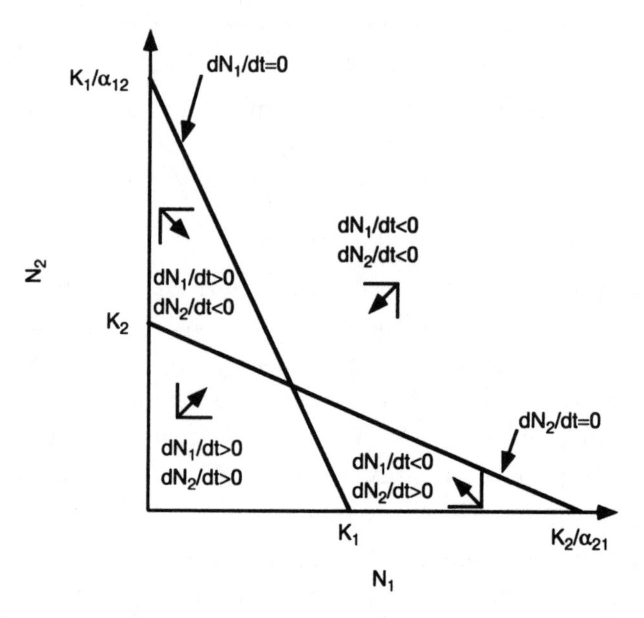

FIGURE 7.5. Phase plane for competition in which both species coexist. The phase plane in Figure 7.4 is completed by drawing in the direction of the change in species numbers in the four regions of the phase plane created by the isoclines. From the arrows indicating change in species numbers, one can see that the system ends up at the interior equilibrium.

$$K_2 < \frac{K_1}{\alpha_{12}}. \tag{7.11}$$

If the necessary condition for coexistence is violated, we know that one of the species will be eliminated.

These conditions involve both the carrying capacities and the competition coefficients. From this pair we can derive a necessary (but not sufficient) condition for coexistence in terms of the competition coefficients alone. Rearrange the conditions for coexistence as

$$\frac{K_1}{K_2} < \frac{1}{\alpha_{21}} \tag{7.12}$$

$$\alpha_{12} < \frac{K_1}{K_2}, \tag{7.13}$$

so the condition for coexistence becomes

$$\alpha_{12} < \frac{K_1}{K_2} < \frac{1}{\alpha_{21}}. \tag{7.14}$$

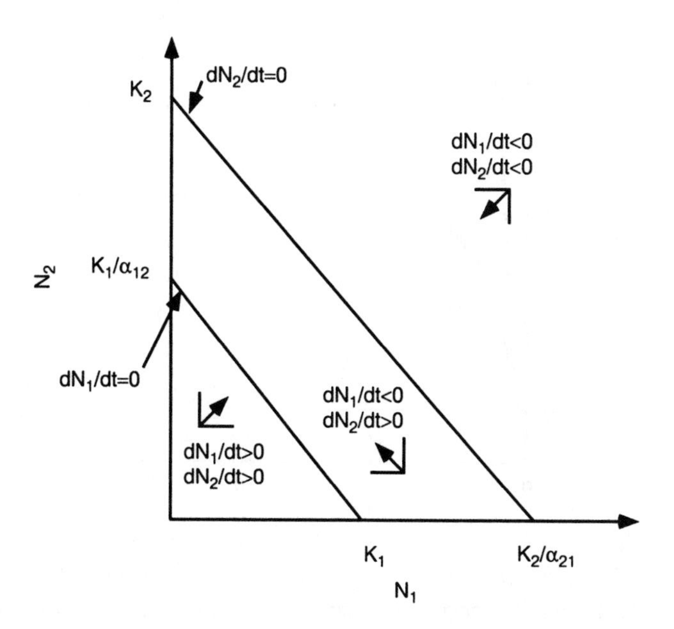

FIGURE 7.6. Phase plane for competition in which species 2 outcompetes species 1. The phase plane is changed from that in Figure 7.5 by changing the relative position of the isoclines.

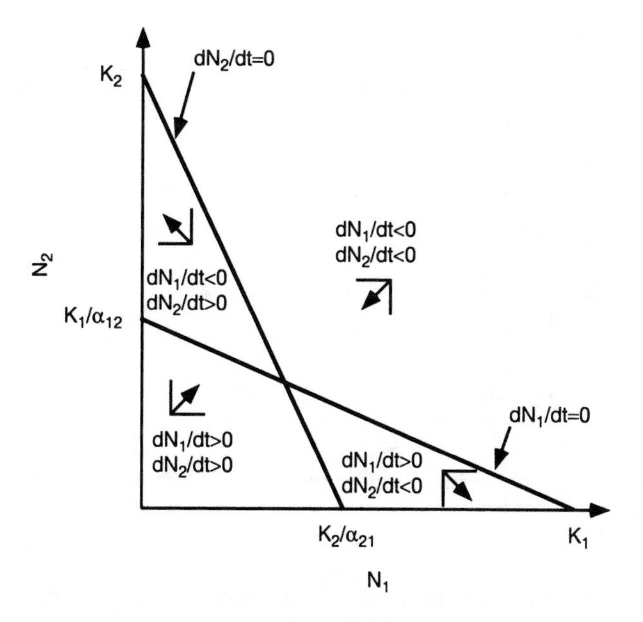

FIGURE 7.7. Phase plane for competition in which the outcome depends on the initial conditions – which species is eliminated depends on the initial conditions.

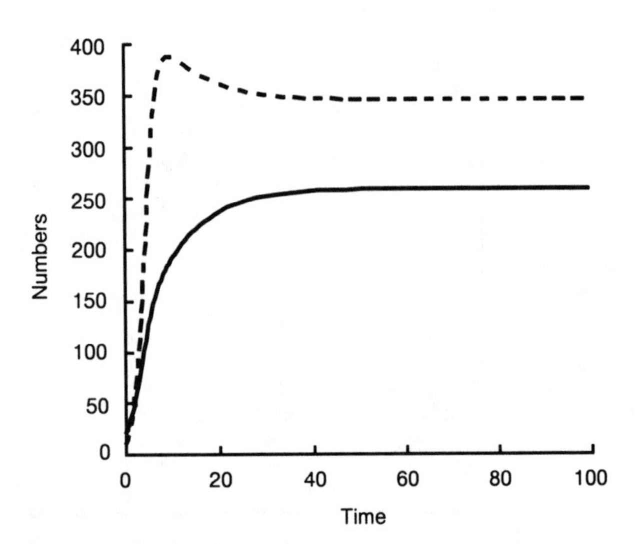

FIGURE 7.8. Dynamics of competition with coexistence. The curves represent a solution of the Lotka–Volterra competition model with parameter values $r_1 = 0.9$, $r_2 = 0.5$, $\alpha_{12} = 0.6$, $\alpha_{21} = 0.7$, and $K_1 = K_2 = 500$. Observe that the conditions for coexistence are satisfied, and that at the joint equilibrium the sum of the population sizes of the two species is greater than 500.

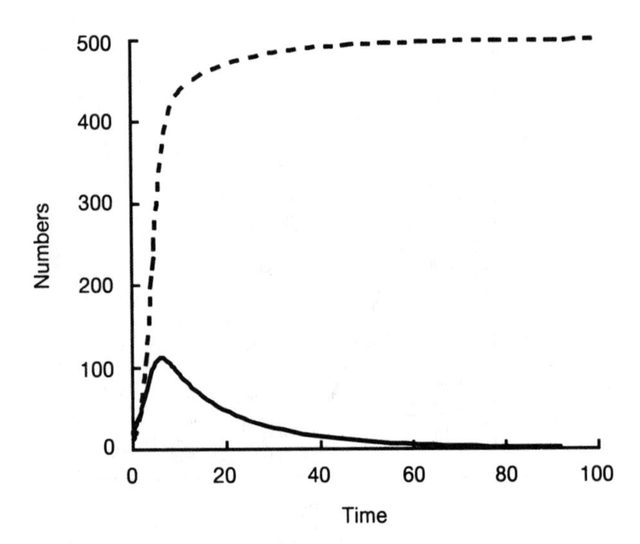

FIGURE 7.9. Dynamics of competition with one species outcompeting the other. The curves represent a solution of the Lotka–Volterra competition model with parameter values $r_1 = 0.9$, $r_2 = 0.5$, $\alpha_{12} = 0.6$, $\alpha_{21} = 1.1$, and $K_1 = K_2 = 500$.

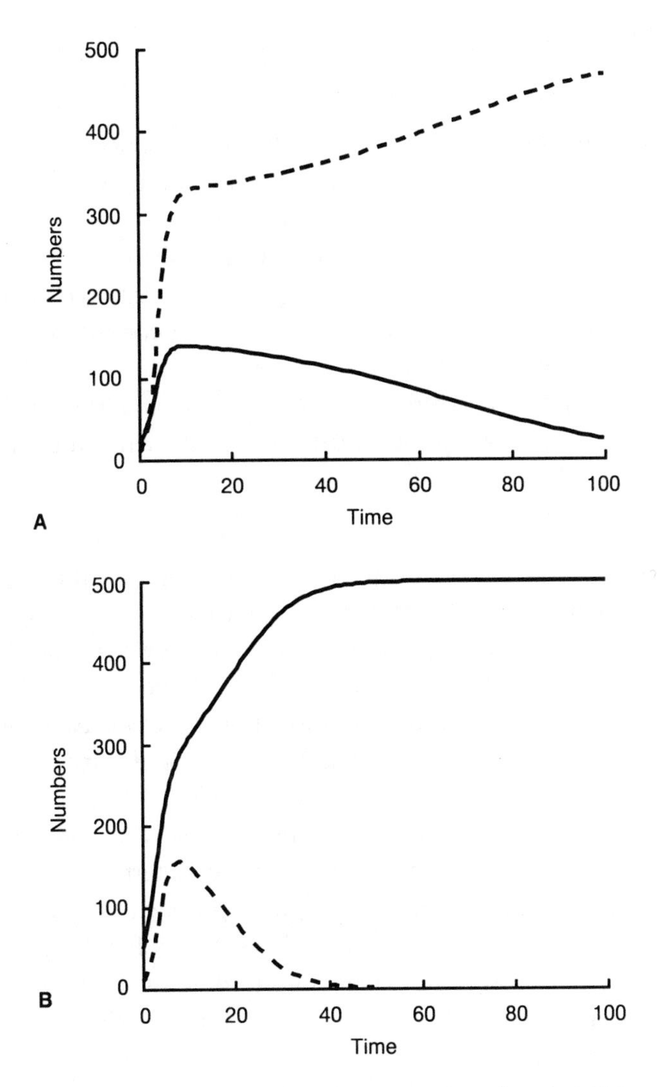

FIGURE 7.10. Dynamics of competition where the outcome depends on the initial conditions. The curves represent a solution of the Lotka–Volterra competition model with parameter values $r_1 = 0.9$, $r_2 = 0.5$, $\alpha_{12} = 1.2$, $\alpha_{21} = 1.1$, and $K_1 = K_2 = 500$. The only difference between the two figures is the initial condition. This is the case outlined in Figure 7.7.

For this condition to be satisfied, we must have

$$\alpha_{12} < \frac{1}{\alpha_{21}} \tag{7.15}$$

or, rearranging,

$$\alpha_{12}\alpha_{21} < 1. \quad (7.16)$$

We interpret this condition as saying that coexistence requires that the geometric mean effect of each species on the other must be less than the effect of each species on itself.

The geometric mean of two numbers is the square root of their product.

If we assume that competition results from the exploitation of a common resource, such as food, then the competition coefficients α_{12} and α_{21} must be equal, so the condition for coexistence is then stated as the principle that each species must affect the other less than itself. This can be rephrased as Gause's *competitive exclusion principle*: to coexist, species must differ in their resource use.

Recall that in the model the effect of each species on itself is 1.

Stability and equilibria

One of the most important outcomes of our graphical analysis of the competition model is that the long-term outcome is always an approach to a stable equilibrium and not to cycles. Thus, if we observe that the dynamics of two species do not eventually lead to a stable equilibrium, then competition between two species, of the form assumed in the Lotka–Volterra model, cannot explain the interaction.

What might lead to an outcome other than a stable equilibrium? One possibility is competition between three or more species. Think of other possible biological situations, or of features of two-species interactions that might lead to other outcomes.

Another approach to analyzing the models would be to solve for the equilibria analytically, and then to determine the stability of the resulting equilibria using the approach outlined in the previous chapter. The outcome is the same as for the graphical analysis. Because we will have to go through a stability analysis in the next chapter, this approach is left for the problems.

7.2 Extensions to Lotka–Volterra models

How can we 'test' the predictions of the Lotka–Volterra models? We observe, both in nature and the laboratory, all possible outcomes – coexistence or elimination of one species or the other – and parameters are difficult to measure. Are there ways to make other tests of the models?

One way to 'test' the Lotka–Volterra model is to take a fixed environment, such as a garden plot for plants or a bottle for *Dro-*

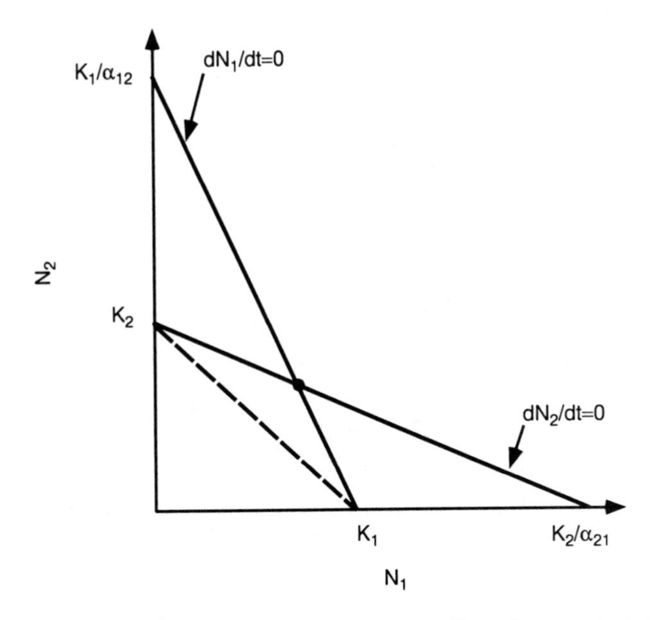

FIGURE 7.11. Relationship of coexistence equilibrium to equilibria of two species in isolation for Lotka–Volterra competition.

sophila, and grow each of two species in isolation and together. The Lotka–Volterra competition model makes a strong qualitative prediction about the relationship among population numbers in these three cases, if the species can coexist. This is illustrated in Figure 7.11. In isolation, species 1 will be at the population level K_1, while species 2 will be at the population level K_2. Note that if the species coexist, the dotted line joining these two points always lies below the coexistence equilibrium. The equation for the dotted line joining these two isolation equilibria is

To see that this is the right equation, look at the point where $N_2 = 0$ and at the point where $N_1 = 0$.

$$\frac{N_2}{K_2} + \frac{N_1}{K_1} = 1. \tag{7.17}$$

As the coexistence equilibrium lies above and to the right of the line 7.17, it must satisfy

$$\frac{N_2}{K_2} + \frac{N_1}{K_1} > 1. \tag{7.18}$$

This prediction, that the joint equilibrium lies above the line joining the two single-species equilibria, can be easily tested by first

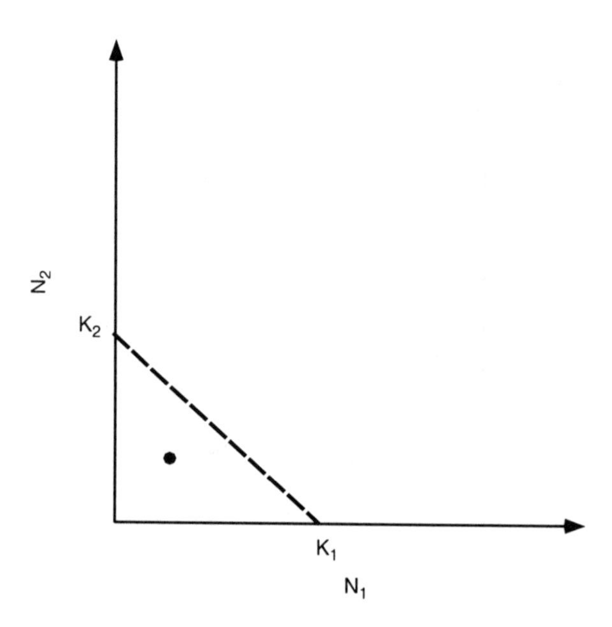

FIGURE 7.12. Relationship of coexistence equilibrium to equilibria of two species of *Drosophila* in isolation as examined by Ayala et al. (1973).

growing two species separately and then together. This was done by Ayala et al. (1973) for *Drosophila pseudoobscura* and *Drosophila serrata*, obtaining the result illustrated schematically in Figure 7.12.

A similar experiment had been performed almost 40 years earlier by Gause (1935) using two species of *Paramecium*, as illustrated in Figure 7.13. Here as well, the position in the phase plane of the coexistence equilibrium contradicts the Lotka–Volterra theory for one of the two experiments performed. The experiments differed in the food sources used for the *Paramecium*.

What is the explanation for the location of the coexistence equilibrium relative to the single-species equilibria? The explanation is that the isoclines must be curved, as illustrated in Figure 7.14. Thus we conclude that even in a system as simple as *Drosophila* or *Paramecium* in a bottle, the Lotka–Volterra model is inadequate.

What are the possible explanations for curved isoclines?

- Frequency-dependent competition, which is not included in the Lotka–Volterra equations, would make the isoclines curved.

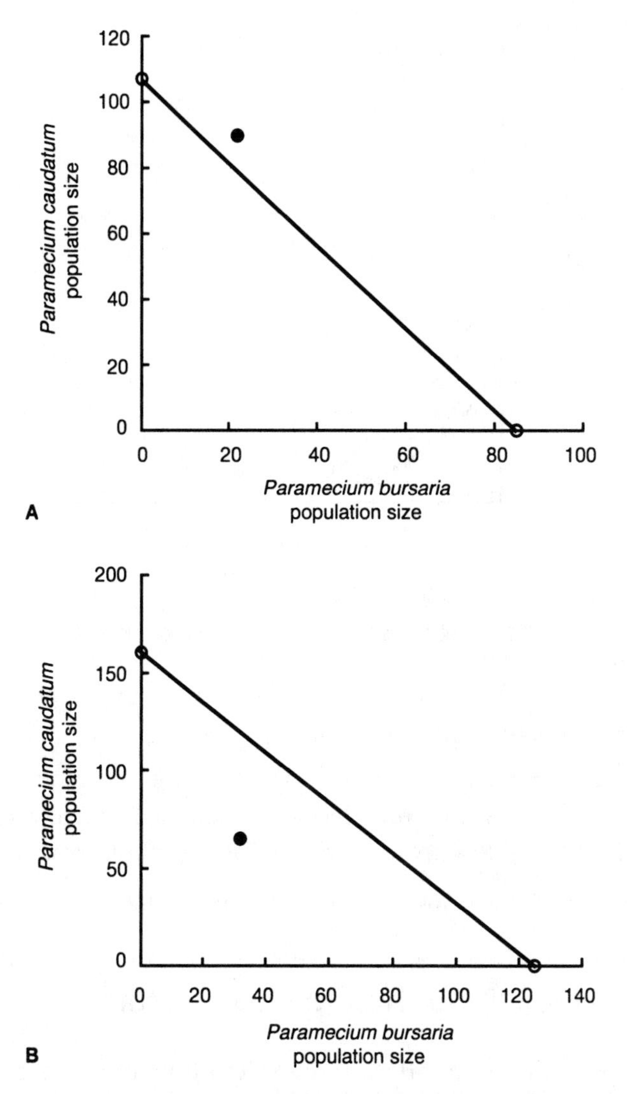

FIGURE 7.13. Outcome of competition between *Paramecium bursaria* and *P. caudatum* using data from Gause (1935). In each case the estimates of carrying capacities and joint equilibrium given by Gause (1935) are used. The open circles represent what are unstable equilibria in the two-species case; the number of each species in isolation. The closed circles represent the stable two-species equilibrium approached in the experiment. The two experiments differed in the strain of bacteria used as food. In the top experiment the joint equilibrium is above the line joining the one-species equilibria, as predicted by the Lotka–Volterra theory. However, in the bottom experiment the joint equilibrium is below the line joining the one-species equilibria, contradicting the Lotka–Volterra theory and showing that even in this simple case more complex theories are needed.

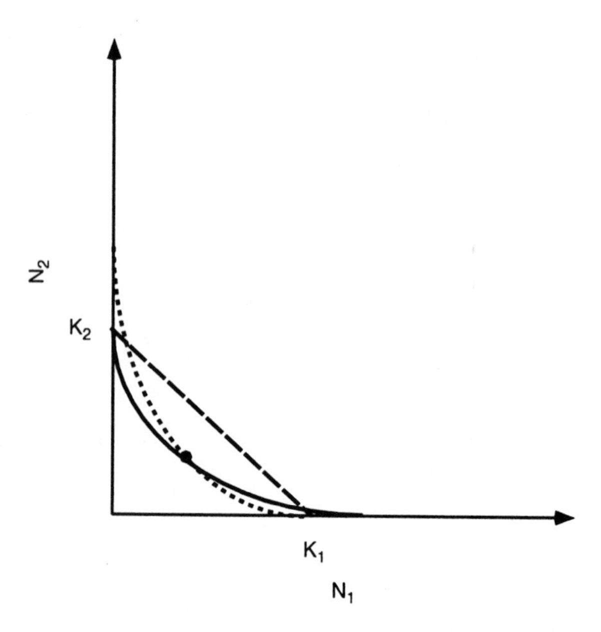

FIGURE 7.14. Curved isoclines in a competition model.

- Energetic considerations would lead to curved isoclines in mechanistic models (Schoener, 1974).
- Age dependence, where larval and adult competition are described by different models, each with linear isoclines, would lead to curved isoclines for the whole system.

We do not pursue these complications further here.

7.3 Competition in field experiments

Is competition an important force shaping natural communities? We have seen how competition can act in model systems, and in laboratory systems, but we have not seen whether competition can be detected in more complex systems. One can look for competition in specific systems, or ask how common competition is in general.

Two surveys, by Schoener (1983) and Connell (1983), both asked essentially the same question of how prevalent competition is in field experiments. Although the details of the two literature surveys are different, both authors concluded that competition was found quite often – more than half the time – in field experiments. Even though both found competition in many studies, the likelihood of finding competition as a function of trophic level was different in the two studies. Schoener did find support for the thesis advanced by HSS, but Connell did not. It is not obvious why the two reviews reached different conclusions.

The HSS hypothesis was discussed at the beginning of Chapter 6.

One illustrative example of an experiment testing competition in a specific system is that performed by Brown and his collaborators (Brown and Davidson, 1977; Brown et al., 1979) to look at competition between ants and rodents in the Arizona desert. In these experiments, if either kind of animal was removed, then the other species showed a dramatic increase in its density. This is strong evidence for competition. There was also a mechanism for competition: the animals all shared the same food source, seeds, in a habitat with limited food supplies.

7.4 Competition for space

Another approach to looking at competition is to consider competition for space. Alternatively, we can think of the effects of spatial structure in a patch model. We will look at the simple case of a competitive hierarchy, which is a simple extension of our metapopulation (patch) model for one species. The possible transitions at one location in the model are illustrated in Figure 7.15. We will assume that species 1 always outcompetes species 2, and that the time scale for this to take place is very short. Thus, species 2 does not change the dynamics of species 1. The dynamics of species 1 therefore are given by an equation which says that the rate of change of the fraction of habitat occupied by species 1 is given by the colonization rate of species 1 minus the extinction rate. The colonization rate of species 1 is proportional to the product of the fraction of patches not occupied by species 1 (those available for colonization) and the fraction of patches occupied by species 1 (and producing colonizers). The extinction rate is simply assumed to be proportional to the fraction of patches occupied by species 1. These assumptions lead to the model equation

The habitat sites occupied by subpopulations of a metapopulation can be called patches. An example of such a site is the host plant of an insect.

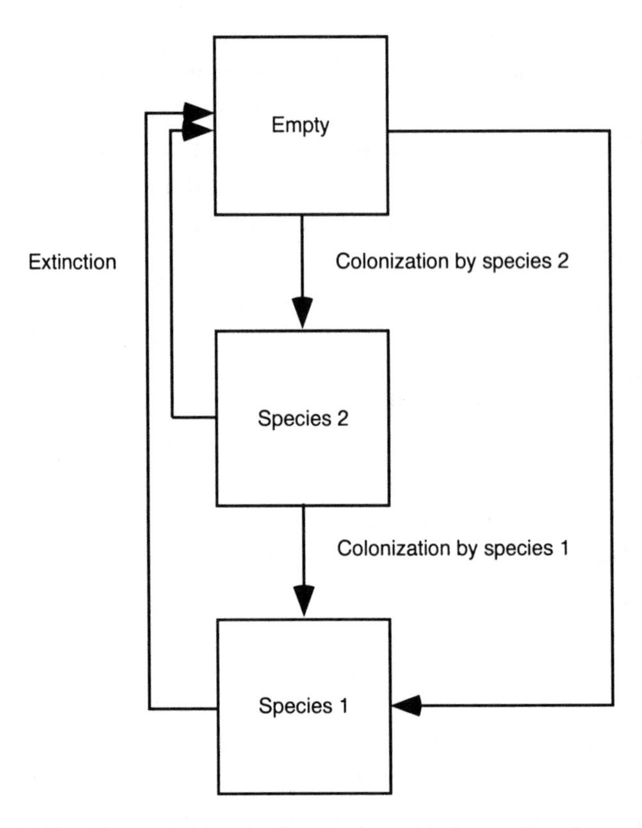

FIGURE 7.15. Transitions at one location in a simple model of competition for space, where species 1 always outcompetes species 2. In the model there is one term for each arrow leading to or from each box.

We do not need the equation for empty space or patches, because the proportion of empty space is just the space not occupied by either species.

$$\frac{dp_1}{dt} = m_1 p_1(1 - p_1) - ep_1, \tag{7.19}$$

where p_1 is the fraction of patches occupied by species 1, m_1 is a measure of the colonization rate of species 1, and e is the extinction rate.

The dynamics of species 2 are derived similarly, except that species 2 cannot colonize a patch occupied by species 1 and that some patches occupied by species 2 are lost as a result of colonization by species 1. Thus, those patches available for colonization are the empty patches, with frequency $1 - p_1 - p_2$. Also, we need to add a loss term accounting for the colonization by species 1 of patches occupied by species 2. This leads to an equation for the dynamics of species 2:

$$\frac{dp_2}{dt} = m_2 p_2(1 - p_1 - p_2) - m_1 p_1 p_2 - e p_2. \qquad (7.20)$$

The analysis of this model is pursued further in the problems. The interesting feature is that the outcome can be coexistence, or species 1 eliminating species 2, or species 2 eliminating species 1. The competitive advantage of species 1 within a patch or at a single site can be offset by the superior dispersal capabilities assumed for species 2. This model has been used quite extensively to understand problems in conservation biology where the effects of habitat destruction are considered.

Summary

We have described three approaches for studying competition. Models can be used to show the kind of dynamics resulting from competition and the conditions that allow species to coexist, or possibly to be eliminated by competition. Cycles cannot occur in competition between two species. In the laboratory, one can show that the process of competition operates, and that even in this simplified setting the Lotka–Volterra competition models are not adequate descriptions. Studies of competition in the field show that competition does occur, but we have not shown that it is *the* force regulating populations.

Problems

1. This problem is based on ideas from Slobodkin (1961, 1964). Slobodkin looked at competition between a brown hydra, *Hydra littoralis*, and a green hydra, *Chlorohydra viridissima*, in laboratory experiments. In these experiments, he was able to achieve coexistence only by the process he called *rarefaction*, removing a fraction of the population of both species at regular intervals by removing part of the medium in which the animals were grown. Demonstrate how this works by adding the term $-mN_i$ to equations (7.3) and

(7.4), representing the effect of the experimenter. If, without the additional term, coexistence is impossible and one species always eliminates the other, show that it is possible to have coexistence with the additional term. Do this by starting with isoclines arranged so that coexistence is impossible, and showing that the additional term could produce equations with isoclines where coexistence is possible.

2. The model we presented of competition for space can be analyzed quite easily. One could analyze the model using phase plane techniques, but we take a different approach that yields additional insights.

   (a) First, find the nonzero equilibrium for the dominant competitor by setting the right-hand side of equation (7.19) equal to zero and solving for p_1. What conditions on the parameters are required for this equilibrium to be positive? Give an ecological interpretation.

   (b) Assuming that the dominant competitor can survive, determine the resulting nonzero equilibrium for species 2 by substituting the equilibrium value for p_1 into equation (7.20), equating the right-hand side to zero, and solving for p_2. What must be true about the colonization rate m_2, relative to the colonization rate m_1, for both species to survive? Does this make ecological sense?

   (c) Start with colonization rates and extinction rates that allow both species to survive. Which species would be eliminated first (at equilibrium) if the extinction rate was slowly increased? (For which species will the equilibrium become negative at the highest value of e?)

   (d) If both species have positive equilibrium levels, what happens to the equilibrium level of species 2, as the extinction rate is increased? Does this make sense?

3. We have not gone through the procedure outlined in Box 6.1 for analytically determining the equilibria and stability of equilibria in the Lotka–Volterra competition model. This is a relatively straightforward problem to solve. Find the equilib-

ria of the Lotka–Volterra competition model and determine their stability. Do the results agree with the phase plane analysis?

4. Discuss the possibilities and difficulties of detecting competition in the laboratory and the field. You will need to refer to some of the suggested reading to complete your answer.
   (a) Describe competition in the laboratory and the field using Lotka Volterra equations.
   (b) Discuss looking at numbers of two species grown together.
   (c) Discuss the approach of looking at overlap in resources used by putative competitors.
   (d) Discuss experimental field manipulations of populations.

Suggestions for further reading

The two surveys on competition in field experiments, by Connell (1983) and Schoener (1983), provide many more details on attempts to discover competition in the field. Two classic studies, Park (1948, 1954) and Connell (1961), are important to read. Park looked at competition in laboratory populations of the flour beetle, *Tribolium*; Connell demonstrated competition in the intertidal zone. Harper and McNaughton (1962) provide a classic example of a study of plant competition for which the Lotka–Volterra approach is not appropriate.

The model describing competition for space is a special case of one first developed in Hastings (1980). The model has been recently used to look at questions in conservation biology and competition for space in plants by Nee and May (1992) and by Tilman (1994).

8

Predator–Prey Interactions

Several long-term data sets have been collected of predator and prey in natural systems with the interaction between lynx, *Lynx canadensis*, and its prey, snowshoe hare, *Lepus americanus* (MacLulick, 1937), providing some of the best long-term data, as illustrated in Figure 8.1. We would like to explain several features of this interaction which are typical of predator–prey relationships. The predator and prey both appear to persist over a long period of time. The coexistence involves cycles: both predator and prey numbers appear to oscillate. Finally, the numbers of predators appear to increase in response to the numbers of prey: the peaks in predator abundance appear to follow the peaks in prey abundance (this is clearest for the last several peaks in the figure).

The presence of oscillations is typical of simple predator–prey interactions, but not typical of competition between two species.

We can reason verbally that if a predator is too efficient it will drive the prey to extinction, and that if the predator is not efficient enough the predator will go extinct. Is there something special about the predator–prey relationships we observe in nature that allows both species to persist? We can also ask if the predator can regulate, or limit the growth of, the prey population. Finally, we ask what are the dynamics of coexisting predator and prey?

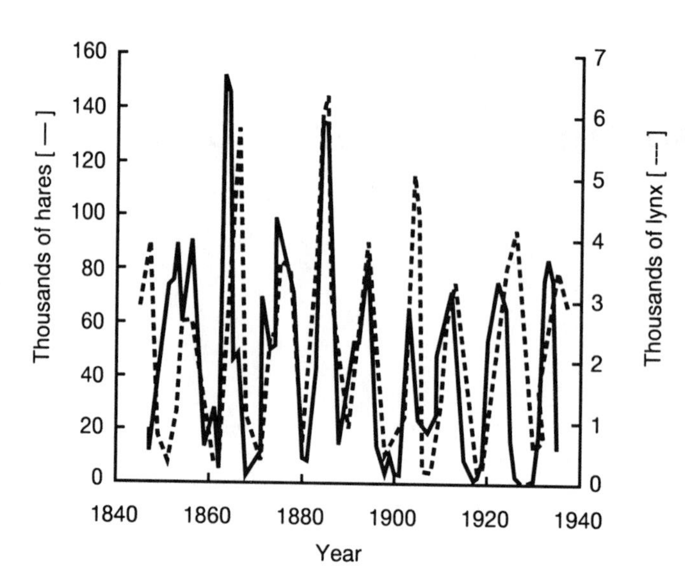

FIGURE 8.1. Dynamics of the predator–prey interaction between lynx (predator) and snowshoe hare (prey) in Canada (data from MacLulick 1937).

8.1 Lotka–Volterra models

Ask yourself about the reasonableness of these assumptions.

We turn to models, just as in our study of competition. We begin with the simplest version of the Lotka–Volterra model, with the following assumptions:

- In the absence of the predator, the prey grows exponentially.
- In the absence of the prey, the predator dies off exponentially.
- The 'per predator rate' at which prey are killed is a linear function of the number of prey, as illustrated in Figure 8.2.
- Each prey death contributes identically to the growth of the predator population.

We now formulate a model according to the assumptions we have just made. Let the number of prey be denoted by H (standing for herbivores) and the number of predators be denoted by P. Our first assumption implies that if $P = 0$ then

$$\frac{dH}{dt} = rH, \tag{8.1}$$

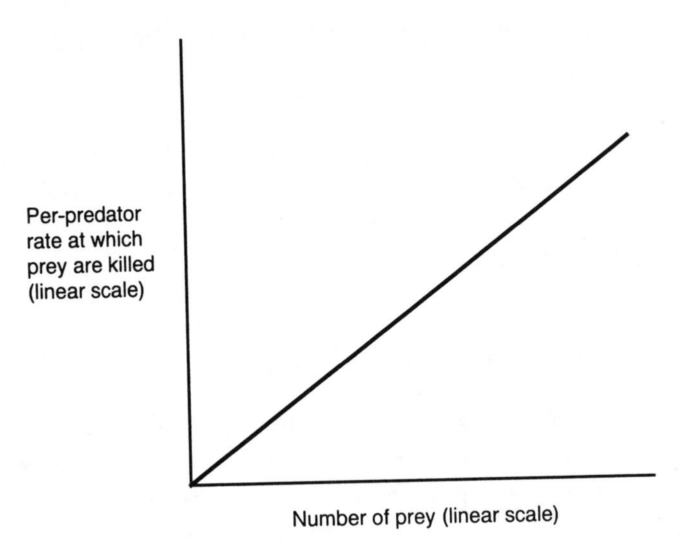

FIGURE 8.2. Predation rate, or functional response, in Lotka–Volterra model.

where r is the intrinsic rate of increase of the prey in the absence of the predator. Our second assumption implies that if $H = 0$ then

$$\frac{dP}{dt} = -kP, \tag{8.2}$$

where k is the rate of decline of the predator in the absence of the prey. Thus $1/k$ is the mean lifetime of the predator. Our third assumption says that the death rate of prey from predation is proportional to the product of prey and predator numbers, bHP, where b is a constant. This says that the full equation for the dynamics of the prey is

To see that $1/k$ is the mean lifetime of the predator compute, $\int_0^\infty e^{-kt}\,dt = 1/k$.

$$\frac{dH}{dt} = rH - bHP. \tag{8.3}$$

Our fourth assumption says that the contribution of predation to the growth rate of the predator population is given by cHP, where c is a constant. Thus, the full equation for the dynamics of the predator is given by

The contribution of predation to the growth of the predator population is directly proportional to the loss of prey from predation that enters into the prey dynamics.

$$\frac{dP}{dt} = cHP - kP. \tag{8.4}$$

8.2 Dynamics of the simple Lotka–Volterra model

Does the model explain the dynamic behavior of a predator–prey system we have seen in Figure 8.1? The only way to answer this question is to determine the dynamics of the model. The first step is to find the equilibria and the stability of the equilibria.

Graphical analysis and linearization

Our first analysis of the model is graphical, as illustrated in Figure 8.3. The determination of the isoclines, and the direction of change of numbers of predators and prey, proceeds exactly as in the competition models analyzed in the previous chapter. Therefore, we do not go through the steps of the analysis, but simply present the results in the figure.

From the graphical analysis, we see that a new feature emerges that was not present in the competition models. There is a tendency for solutions to oscillate. However, we are unable to conclude whether the equilibrium point is stable from the graphical analysis. Although we have indicated a cycle in Figure 8.3, we do not yet know whether solutions spiral toward the equilibrium, truly cycle, or spiral out. We therefore proceed with a stability analysis of equilibrium points, as described in Chapter 6.

Our first step is to find the equilibria of the model. We will let $F(H, P)$ be the growth rate of the prey population and let $G(H, P)$ be the growth rate of the predator population. We find the equilibria of the model by simultaneously solving the pair of equations $F = 0$, $G = 0$. From the equation $F = 0$, we find that

These are just the equations of the isoclines.

$$rH - bHP = 0, \tag{8.5}$$

or

$$H(r - bP) = 0. \tag{8.6}$$

We conclude that either

$$H = 0 \tag{8.7}$$

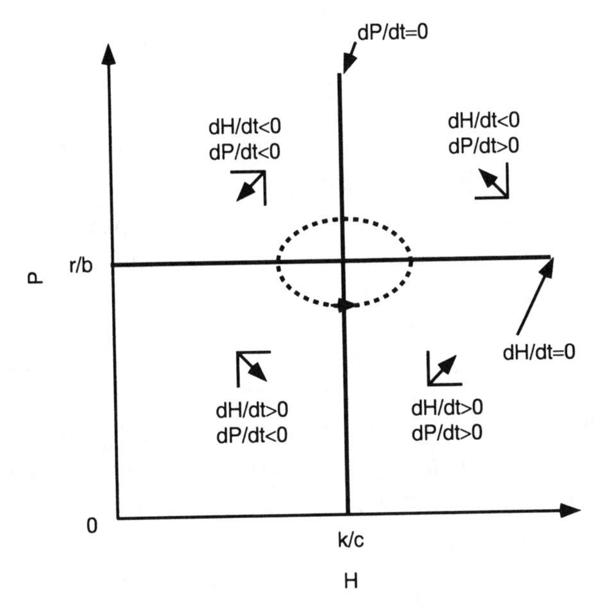

FIGURE 8.3. Phase plane for Lotka–Volterra predator–prey model with no density dependence. The first step in drawing this phase plane is to draw the two isoclines, given by equations (8.8) and (8.12). The next step is to draw the arrows indicating the direction of population change in each of the four sectors created by the isoclines. From this phase plane we are unable to determine the stability of the equilibrium, in contrast to the competition case where stability could be determined from the phase plane. The oscillation indicated on the phase plane can be found by numerical (computer) methods or more sophisticated mathematical methods.

or

$$P = \frac{r}{b}. \tag{8.8}$$

In a similar fashion, we use the equation $G = 0$ to find

$$cHP - kP = 0, \tag{8.9}$$

or

$$P(cH - k) = 0. \tag{8.10}$$

We conclude that either

$$P = 0 \tag{8.11}$$

It is typical of predator–prey models that the equation for the predator dynamics completely determines the equilibrium level of the prey. We will see later that this corresponds to our biological assumption that the per predator predation rate is independent of the number of predators: the predators act independently.

or

$$H = \frac{k}{c}. \tag{8.12}$$

There are two equilibria for the model, one with both species zero, and one where both prey and predator are nonzero. From the graphical analysis, we can conclude that the equlibrium with both species absent is unstable. We thus concentrate our stability analysis on the nontrivial equilibrium

Translate this statement about stability of equilibria with species absent into biological terms. For example, if both species are absent, and we introduce a few of the prey species, what happens?

$$(H, P) = (\frac{k}{c}, \frac{r}{b}). \tag{8.13}$$

Our first step in determining the stability of this equilibrium is to approximate the dynamics near the equilibrium by linearizing, as outlined in Box 6.1. We find that the linearization of this model near the equilibrium is given by

We use the symbol J for the matrix, because this linearization matrix is known mathematically as a Jacobian matrix. This is the community matrix we discussed in Chapter 6.

$$J = \left(\begin{array}{cc} \frac{\partial F}{\partial H} & \frac{\partial F}{\partial P} \\ \frac{\partial G}{\partial H} & \frac{\partial G}{\partial P} \end{array} \right) \Bigg|_{(H,P)=(\frac{k}{c},\frac{r}{b})} \tag{8.14}$$

$$= \left(\begin{array}{cc} r - bP & -bH \\ cP & cH - k \end{array} \right) \Bigg|_{(H,P)=(\frac{k}{c},\frac{r}{b})} \tag{8.15}$$

$$= \left(\begin{array}{cc} r - b\frac{r}{b} & -b\frac{k}{c} \\ c\frac{r}{b} & c\frac{k}{c} - k \end{array} \right) \tag{8.16}$$

$$= \left(\begin{array}{cc} 0 & -b\frac{k}{c} \\ c\frac{r}{b} & 0 \end{array} \right) \tag{8.17}$$

Complex eigenvalues and oscillations

We will now find the eigenvalues of the matrix J to determine the behavior of the model near equilibrium. Mathematically this is analogous to the procedure we used in analyzing age-dependent growth. Thus we set the determinant of $J - \lambda I$ to be zero:

$$0 = \left| \begin{array}{cc} -\lambda & -b\frac{k}{c} \\ c\frac{r}{b} & -\lambda \end{array} \right| = \lambda^2 + rk \tag{8.18}$$

The solution to this equation requires the use of complex numbers (see Box 8.1). We conclude that

$$\lambda = \pm \imath \sqrt{rk}. \tag{8.19}$$

Thus in this model the equilibrium point is neither locally asymptotically stable (solutions do not get closer) nor unstable, as determined by the linear approximation. A more sophisticated analysis of this model shows that in fact all the solutions are closed curves, as illustrated in Figure 8.3. Thus we conclude that we must make this model more realistic before we can begin to draw biological conclusions about stability. However, the information about the period of the oscillations can prove useful.

Before we leave our study of this model, we will indicate how we could have determined some of the qualitative information about the dynamics from an examination of the phase plane. As we noted earlier, the entries in the linearization of a predator–prey model take the form

$$\begin{pmatrix} ? & - \\ + & 0 \end{pmatrix}. \tag{8.20}$$

Remind yourself again why the entries take the signs indicated. The '?' in the upper left arises because the effect of the prey species on itself can be affected by the predation term, as we will see.

We will explain how to read off the signs of both the entries filled in and the entries not filled in from the phase plane. Start with the one in the upper right, which is $\partial F/\partial P$, *evaluated at equilibrium*. This means that we need to look at the effect of changing P on F at equilibrium, while holding H constant. In the phase plane diagram, if we look at what happens, F changes from positive to negative, as indicated in Figure 8.4. Follow the bold arrow and it goes from a region of phase space where F is positive to one where F is negative. Therefore, the partial derivative $\partial F/\partial P$, evaluated at equilibrium, is negative.

In a similar fashion, in the same figure, we note that along the bold arrow G does not change its value: it is always zero. We thus conclude that, at equilibrium, $\partial G/\partial P = 0$. Similarly, we can determine that, at equilibrium, $\partial F/\partial H = 0$ and $\partial G/\partial H > 0$.

Do this yourself graphically – which way should you draw a bold arrow? How should you label the head and tail of the arrow?

Thus, we have concluded that the signs of the entries in the linearization about the equilibrium are

$$\begin{pmatrix} 0 & - \\ + & 0 \end{pmatrix}. \tag{8.21}$$

As the trace is zero, the real part of one eigenvalue is greater than or equal to zero, and the real part of the other eigenvalue must be less than or equal to zero. Since the determinant is positive, we

Box 8.1. Facts about complex numbers and complex eigenvalues.

Complex numbers arise naturally in the determination of eigenvalues for the linearization of predator–prey models, or in systems with age structure and more than two age classes. We collect a number of facts and implications here. We denote the square root of -1 by $\imath$:

$$\imath = \sqrt{-1}.$$

This is called an *imaginary* number, and a *complex* number is just the sum of an imaginary number and a real number. When an eigenvalue of the linearization of a continuous time model involves $\imath$, the eigenvalue is representing a growth rate of a perturbation away from equilibrium. Thus the dynamics near equilibrium are given by an eigenvector multiplied by a term of the form $e^{\lambda t}$, where λ is complex. Let us write

$$\lambda = u + v\imath.$$

Then

$$e^{\lambda t} = e^{ut + \imath vt} = e^{ut} e^{\imath vt}. \tag{a}$$

We understand the exponential e^{ut} on the right-hand side of (a), but to understand the exponential $e^{ut}e^{\imath vt}$ we need to use the important formulas

$$\begin{aligned} e^{\imath\theta} &= \cos(\theta) + \imath\sin(\theta) \\ e^{-\imath\theta} &= \cos(\theta) - \imath\sin(\theta). \end{aligned}$$

Thus, complex eigenvalues correspond to oscillatory behavior. Note that if λ is a complex eigenvalue, λ is always part of a complex pair, $u \pm v\imath$. Consequently, there are two growth rates of deviations away from equilibrium:

$$e^{\lambda t} = e^{ut \pm \imath vt} = e^{ut}\left[\cos(vt) \pm \imath\sin(vt)\right]. \tag{b}$$

Because the oscillatory term shows no long-term change in magnitude, the equilibrium is stable if $u < 0$ and unstable if $u > 0$. (Here u is the real part of the eigenvalue.) The

Box 8.1 (cont.)

period of the oscillatory behavior is given by $2\pi/v$. To get real values for the growth of the perturbations, we can add the two possible solutions given on the right of (b) and divide by 2 to get $e^{ut}cos(vt)$ or take the difference and divide by $2\imath$ to get $e^{ut}sin(vt)$. In each case, the other trigonometric term cancels.

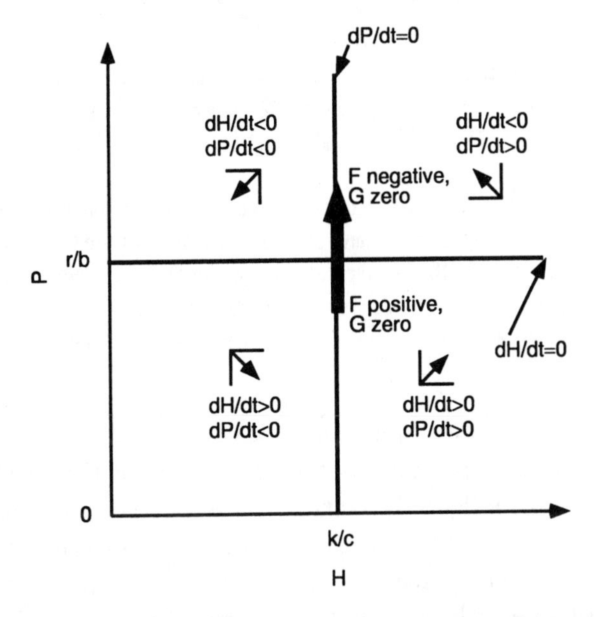

FIGURE 8.4. Determining graphically the linearization at equilibrium of the Lotka–Volterra predator–prey model without density dependence. The bold arrow indicates the meaning in the phase plane of a partial derivative with respect to P. By comparing the sign of F and G at the tail and head of the bold arrow, we can find the sign of $\partial F/\partial P$ and $\partial G/\partial P$. Because F changes from positive to negative, $\partial F/\partial P < 0$. Because G does not change, $\partial G/\partial P = 0$.

see that both eigenvalues must in fact be purely imaginary with real part zero.

Behavior of the Lotka–Volterra model far from equilibrium

This analysis has shown that we cannot determine the stability of the equilibrium using the linearization technique: the model produces an equilibrium that lies right on the border between

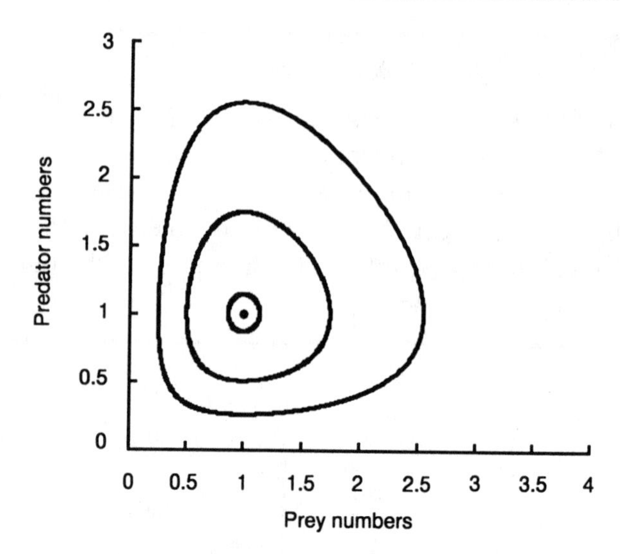

FIGURE 8.5. Numerical solutions of the Lotka–Volterra predator–prey model with no density dependence, displayed in the phase plane. Equations (8.3) and (8.4) are solved numerically with $r = b = c = k = 1$. The closed curves represent sustained oscillations. However, as there is an infinite family of closed curves corresponding to different initial conditions, this model is biologically unrealistic because the outcome will be changed by small changes in the model.

stability and instability. More sophisticated analyses of this model, or numerical solutions, demonstrate that this inability to determine stability actually reflects the fact that the solutions to the full model are always cycles, with *the amplitude of the cycle determined solely by initial conditions* (see Figure 8.5). If the system starts with a large-amplitude cycle, it continues to cycle with large amplitude. If the system starts with a small-amplitude cycle, it continues to cycle with small amplitude. There is no tendency for the amplitude of the cycles to change at all. We have duplicated one feature of the observed dynamics, the presence of oscillations. However, we have no way to explain oscillations of any particular size.

8.3 Role of density dependence in the prey

We argue that the model we have just developed is a poor one to use to describe ecological systems because small changes in the model might lead to large changes in the dynamics of the model. Because the model simply oscillates with whatever amplitude is determined by the initial conditions, a small change in our assumptions might lead to either oscillations growing without bound, or to an approach to equilibrium. Because essentially all models in ecology are rather crude, such sensitive behavior of the model outcome to changes in the underlying model is clearly both undesirable and unrealistic. We thus begin to look at modifications of the basic model, which will serve both to remove this curious and unrealistic behavior and to justify the argument made in this paragraph.

Because the 'pure' Lotka–Volterra model is at the border between stability and instability, we can add various modifications and see if they lead to stability or instability of the equilibrium, and then classify the modification as stabilizing or destabilizing.

One of the simplest modifications possible to the basic Lotka–Volterra model is to change our assumption that the prey grows exponentially in the absence of the predator to say that the prey grows logistically in the absence of the predator. In this case, we find that the equations for our model become

$$\frac{dH}{dt} = rH\left[1 - \frac{H}{K}\right] - bHP \qquad (8.22)$$

and

Adding density dependence does not change the equation for the predator population.

$$\frac{dP}{dt} = cHP - kP. \qquad (8.23)$$

A phase plane analysis of this model is displayed in Figure 8.6. The first step in drawing the phase plane is to find the isoclines. The prey isocline is found by solving

$$0 = rH\left[1 - \frac{H}{K}\right] - bHP, \qquad (8.24)$$

yielding

$$H = 0, \qquad (8.25)$$

or

$$0 = r\left[1 - \frac{H}{K}\right] - bP. \qquad (8.26)$$

As in the competition case, it is easy to draw this line on the phase plane by noticing that it passes though the P axis at $P = r/b$ and through the H axis at $H = K$. The predator isocline is unchanged from the case without density dependence.

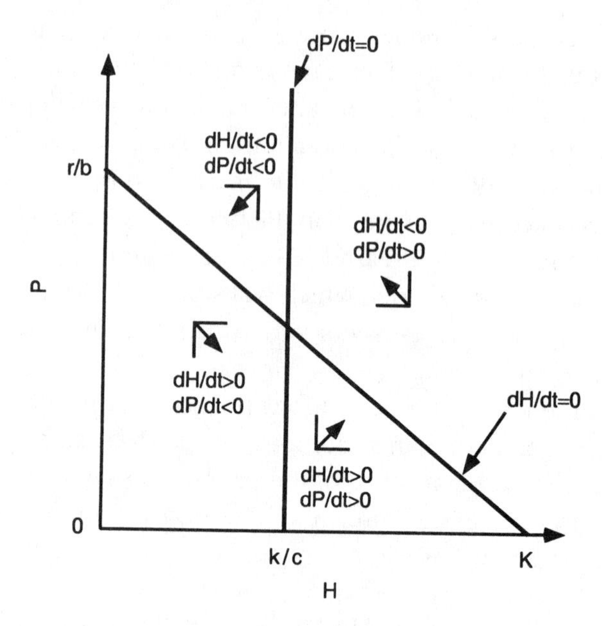

FIGURE 8.6. Phase plane for Lotka–Volterra predator–prey model with density dependence in the prey.

Once again, we determine the signs of the entries in the linearization about the nontrivial equilibrium by following along the bold arrows in Figure 8.7.

Work out for yourself that the signs of the entries in the linearization matrix now have the form

$$\begin{pmatrix} - & - \\ + & 0 \end{pmatrix}. \tag{8.27}$$

The trace is the sum of a negative number and zero, which is negative. The determinant is given by a number that is (−)(0) − (−)(+), which is positive. Thus by the criterion for stability outlined in Box 6.2, we conclude that the equilibrium is stabilized by the addition of density dependence in the prey. However, the equilibrium is still approached in an oscillatory fashion, as illustrated in Figure 8.8. Also, although the equilibrium is stable, the numbers of predators can still reach very low levels before the system reaches equilibrium.

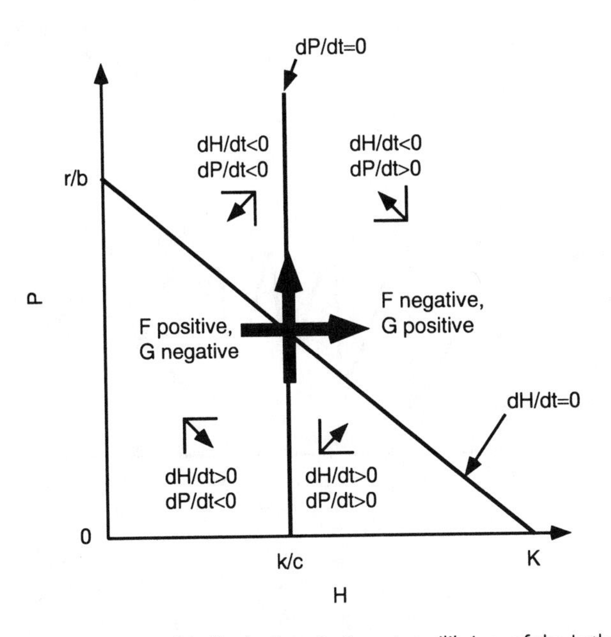

FIGURE 8.7. Determining graphically the linearization at equilibrium of the Lotka–Volterra predator–prey model with density dependence. The horizontal and vertical bold arrows indicate in the phase plane the meaning of the partial derivatives with respect to H and P, respectively. By comparing the sign of F and G at the tail and head of the horizontal bold arrow, we find that $\partial F/\partial H < 0$ and $\partial G/\partial H > 0$. Using the vertical bold arrow, we can find $\partial F/\partial P$ and $\partial G/\partial P$ as we did in the case without density dependence.

8.4 Classic laboratory experiments on predation

How realistic is this conclusion that we obtain stability in predator–prey systems? We will first look at this question in the context of laboratory experiments, and then focus on examples from the field after refining our theory.

Gause

By working with microorganisms, Gause was able to carry out his experiments over a relatively short time and in a relatively small space.

Inspired by the work of Lotka and Volterra on models describing the interaction between predator and prey, Gause (1934) undertook a series of experiments in a microcosm to test the predictions of the theory. In the microcosm, bacteria were supplied as food for *Paramecium caudatum*, which in turn was consumed by *Didinium nasutum*. *Didinium* is a voracious predator, consum-

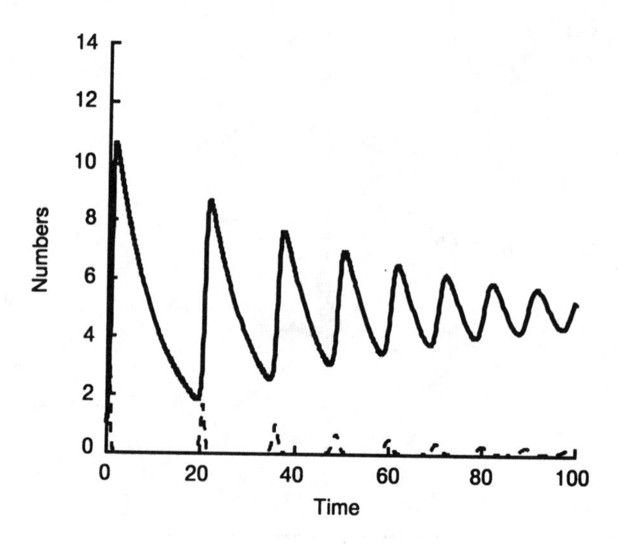

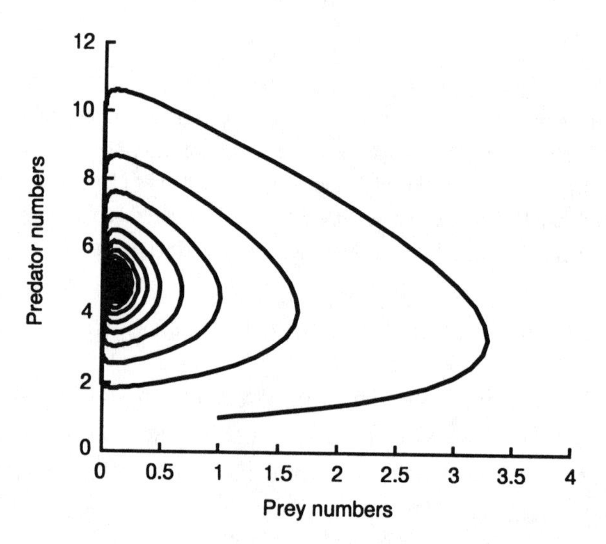

FIGURE 8.8. Dynamics against time (top) and in the phase plane (bottom) for the Lotka–Volterra predator–prey model with density dependence in the prey. In the plot of numbers against time, the prey is the dashed line; the predator is the solid line. Note how close solutions come to the axes before approaching the equilibrium.

ing at least one fresh *Paramecium* every 3 hours. In this system, Gause concentrated on the interaction between *Paramecium* and *Didinium*, with the bacteria introduced as part of the substrate.

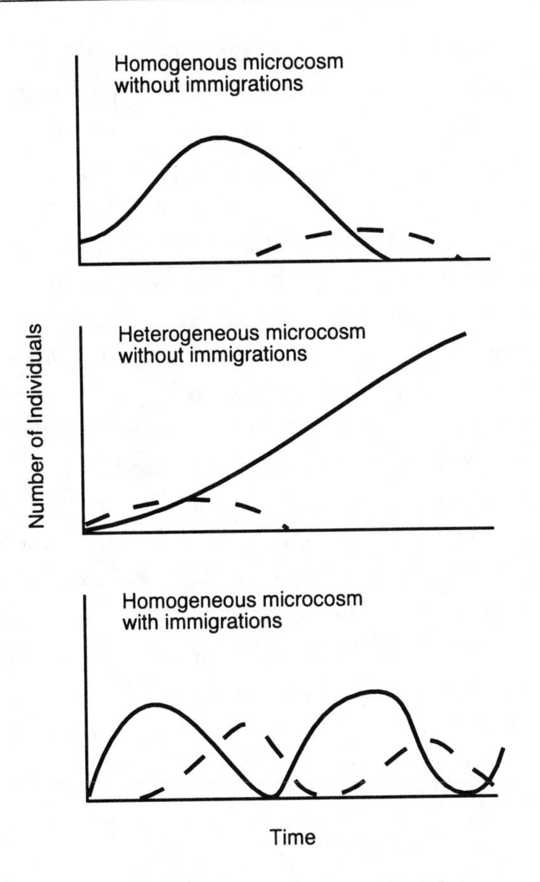

FIGURE 8.9. Schematic diagram of dynamics against time for *Didinium nasutum*, predator, shown in dashed lines, and *Paramecium caudatum*, prey, shown in solid lines, under three different experimental protocols (redrawn from Gause 1935).

In the first experiments performed by Gause, he simply placed the prey, and then the predator, in the experimental microcosm. The outcome was always the same – the predator invariably consumed all the prey, and then the predator also went extinct (see top panel of Figure 8.9). This was not the outcome predicted by the theory, where coexistence was the expected outcome. However, at least the inherent tendency to oscillate did show up in this experiment.

The next step was to try and look for processes that would allow predator and prey to both persist. So Gause tried to provide a refuge for the prey, a place where the *Paramecium* could not be consumed. Once again, the species did not coexist, although the outcome was different (middle panel of Figure 8.9). Here, the predator starved, and those prey protected in the refuge eventually produced exponential growth of the prey population.

Finally, Gause sought a way to produce sustained oscillations and coexistence, which was apparently achieved as illustrated for the case illustrated in the bottom panel of Figure 8.9. However, the only way that this coexistence was achieved was for the experimenter to add 1 *Paramecium* and 1 *Didinium* to the experimental microcosm at regular intervals, thus preventing either species from going extinct. This experiment did demonstrate, however, oscillations in a predator prey system.

One argument as to why predator–prey systems persist in nature is that natural predators typically have many alternate prey species, thus allowing persistence.

At this point the agreement between theory and experiment is quite weak. We will go back to the theory and determine what destabilizing influences we have left out. Before returning to the theory, we first present the results of another classic experiment that built upon the stabilizing influence that Gause used. If persistence of predator and prey could be achieved by the experimentalist adding predator and prey to a microcosm, could the same effect be achieved by predator and prey themselves moving among different habitats?

Huffaker

Huffaker (1958) performed several experiments with a predator–prey system where the habitat was a series of oranges laid out in trays. The predatory mite *Typhlodromus occidentalis* fed upon the six-spotted mite, *Eotetranychus sexmaculatus*, which fed upon the oranges. We describe in more detail one of these experiments, which demonstrated how persistence could be achieved.

In this experiment, the 'universe' consisted of three trays, each containing 40 oranges arrayed in a grid of 4 rows with 10 oranges. Partial barriers were set up between the oranges within a tray, and connections were made between trays. Huffaker then began the experiment by placing 1 female six-spotted mite on each of the oranges. Five days later 27 predators were added to 27 oranges distributed throughout the experimental universe. Af-

Mites crawl from orange to orange.

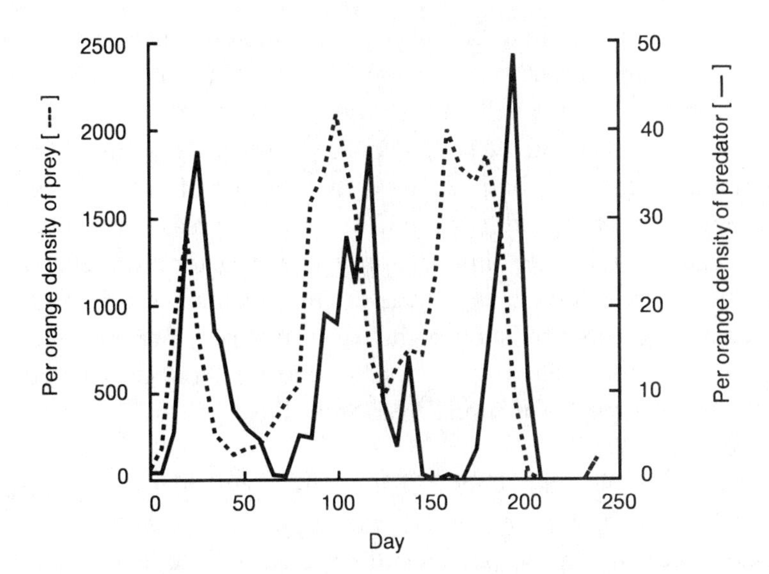

FIGURE 8.10. Oscillations in a laboratory predator–prey system with spatial structure, where the prey, the six-spotted mite *Eotetranychus sexmaculatus*, was eaten by the predator, *Typhlodromus occidentalis*. The experimental universe consisted of 120 oranges (data from Huffaker 1958).

ter this, the numbers of predator and prey were recorded, as well as the locations of 'occupied' oranges.

The results for the total number of predator and prey were quite dramatic, with sustained oscillations resulting (Figure 8.10). We thus have experimental evidence that a metapopulation structure can lead to persistence that would be impossible without this kind of spatial structure. Huffaker had previously shown, not surprisingly, that at smaller spatial scales, persistence did not result.

We have already seen metapopulations in both the context of single species and competing species.

Before presenting further laboratory or field examples, we will return to the theoretical development and try to explain the two classic experiments we have just described.

8.5 Functional response

Gause's (1934) experiments clearly indicated that the Lotka–Volterra model with density dependence in the prey was much more stable than the dynamics of the predator–prey system observed

Think of potentially destabilizing influences.

in the laboratory. Therefore, we conclude that we have left out of the model important destabilizing features of the predator–prey interaction. Here we will analyze the effects of changing another of the assumptions in the Lotka–Volterra predator–prey model that is unrealistic. We recognize that as the number of prey increases the rate of prey capture per predator cannot increase indefinitely. At some point the rate of prey capture per predator must level off, as the number of prey is not the limiting factor. We refer to the predation rate as a function of the number of prey per predator as the *functional response*. If we included just this effect, we would obtain the type II functional response illustrated in Figure 8.11 (Holling, 1959).

Another biological effect is that when a particular prey item is rare, it may be ignored by the predator. This leads to a faster than linear rise in the per predator predation rate as a function of prey numbers, when prey numbers are small. This is a type III functional response, which is also illustrated in Figure 8.11.

Effect of functional response on stability

We will determine the consequences of a type II functional response by determining the effect on stability. We write the model in this case as

$$\frac{dH}{dt} = rH\left[1 - \frac{H}{K}\right] - bf(H)P \tag{8.28}$$

and

$$\frac{dP}{dt} = cf(H)P - kP, \tag{8.29}$$

where the function $f(H)$ represents the functional response. Once again the predator isocline is vertical.

We begin by finding the prey isocline, which is given by

$$rH\left[1 - \frac{H}{K}\right] - bf(H)P = 0. \tag{8.30}$$

We solve for P because it is easier than solving for H.

Solve this for P, obtaining

$$P = \frac{rH(1 - H/K)}{bf(H)}. \tag{8.31}$$

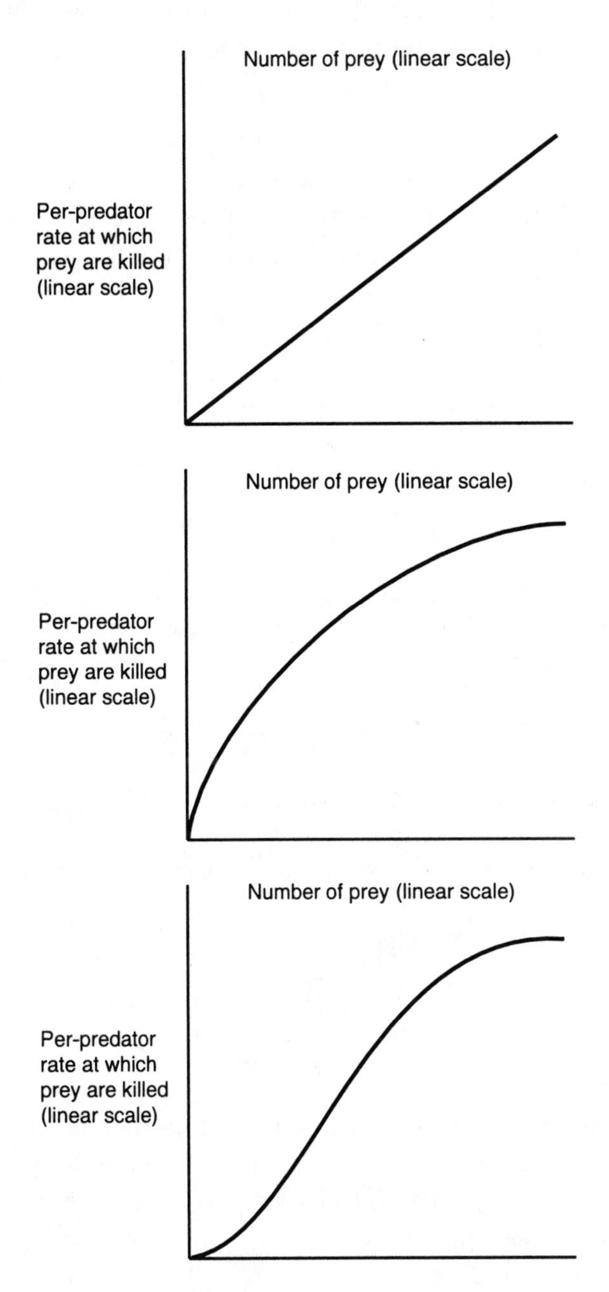

FIGURE 8.11. (Top) Lotka–Volterra or linear functional response. (Middle) Type II functional response. (Bottom) Type III functional response.

We now specialize to a particular form of f that has often been used to model a type II functional response:

$$f(H) = \frac{aH}{1 + dH}, \tag{8.32}$$

If H is large, f is approximately a/d. If H is small, f is approximately aH.

where a and d are positive parameters.

With this choice of f, we can now find the location of the predator isocline. It is given by the solution of the equation

$$0 = \frac{acPH}{1 + dH} - kP, \tag{8.33}$$

which is

$$P = 0, \tag{8.34}$$

or

$$0 = \frac{acH}{1 + dH} - k. \tag{8.35}$$

We can solve (8.35) for H to find the location of the vertical predator isocline

$$H = \frac{k}{ac - kd}. \tag{8.36}$$

This isocline is drawn on Figures 8.12 and 8.13.

We now continue with our computation of the prey isocline for this particular functional response. Substituting from (8.32) into (8.31), we obtain

$$P = \frac{rH(1 - H/K)}{b\left(\frac{aH}{1+dH}\right)} \tag{8.37}$$

$$= \frac{r(1 - H/K)(1 + dH)}{ab}. \tag{8.38}$$

Equation 8.38 is a parabola because the right-hand side is quadratic in H.

To find the points where the parabola crosses the axis, set the numerator of (8.38) equal to zero and solve for H. This is easy to do because the numerator is already factored.

This is a parabola. As we shall see, the important feature will be the location of the maximum, which is located midway between the points where the parabola crosses the H axis (i.e., where P is zero). Since these points are $H = -1/d$ and $H = K$, the maximum is at $H = (K - 1/d)/2$.

We graph two different possibilities in Figures 8.12 and 8.13, depending on the relative location of the maximum (or hump) in the prey isocline and the predator isocline. In the first case,

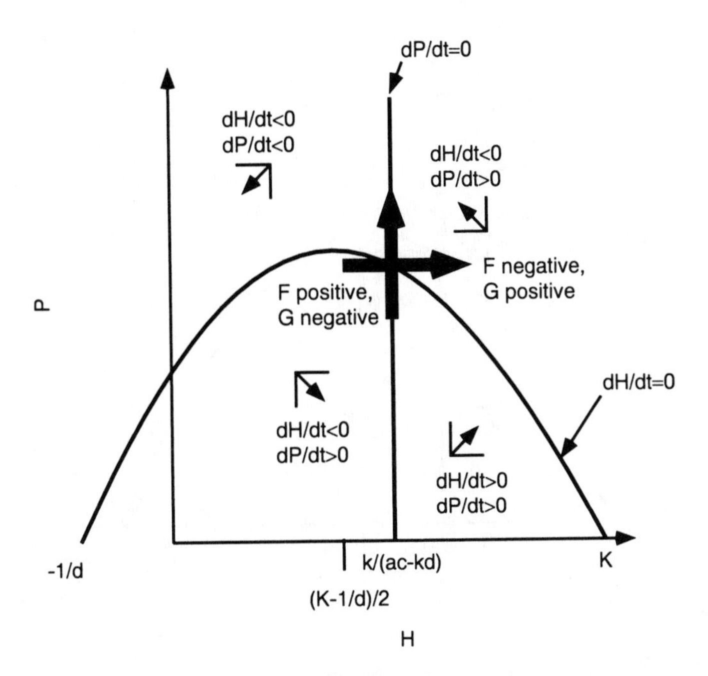

FIGURE 8.12. Phase plane for predator–prey model with type II functional response, stable case. Note that the phase plane near the equilibrium in this case looks like the phase plane depicted in Figure 8.7. The bold arrows in the figure are used to determine the signs of the entries in the linearization (Jacobian) matrix as in Figure 8.7.

where the predator isocline is to the right of the hump (Figure 8.12), the equilibrium point is stable, because in the vicinity of the equilibrium the dynamics are the same as in the case of the Lotka–Volterra response we have already analyzed.

However, if the predator isocline is to the left of the hump, as in Figure 8.13, we have a different situation. We can again determine the signs of the entries in the linearization by following along the bold arrows in the figure, obtaining

$$\begin{pmatrix} + & - \\ + & 0 \end{pmatrix}. \tag{8.39}$$

Now the trace is the sum of a positive number and zero, which is positive. Thus by the criterion for stability outlined in Box 6.2, we conclude that the equilibrium is unstable.

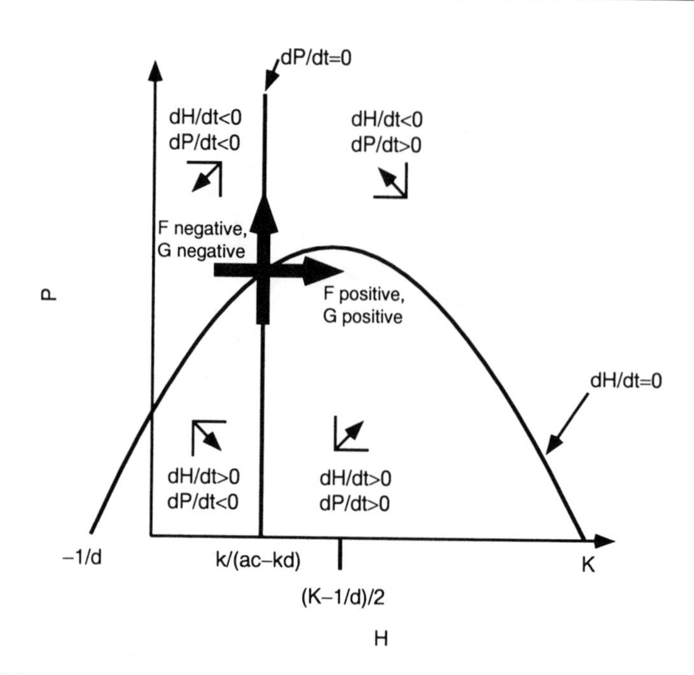

FIGURE 8.13. Phase plane for predator–prey model with type II functional response, unstable case. Contrast the phase plane near equilibrium to the stable case, Figure 8.12.

Ecological implications

We can now discuss what biological changes can cause the system to shift between the stable and unstable cases we have just analyzed. Before doing this we look at the dynamics in the unstable case, which is most easily done by solving the model (8.28)-(8.29) numerically (Figure 8.14). What is important to note from the figure is that in the case where the equilibrium is unstable, the numbers of both species can become very small, so small that if we included any stochastic effects we would expect that one or both species would go extinct. Thus, the inclusion of the functional response produces a model that we can use to explain the observations Gause (1934) made on simple predator–prey systems in the laboratory.

Our deterministic model can never produce the extinction of either species. We need to recognize that when the numbers of either predator or prey become small, we must use intuition and consider that a very small population size in the model may really correspond to extinction if stochastic forces are included.

We have seen that the equilibrium of the predator–prey model, when functional response is included, can be either stable or unstable. What are the biological features which lead to stability or instability? The stability of the equilibrium is determined by the

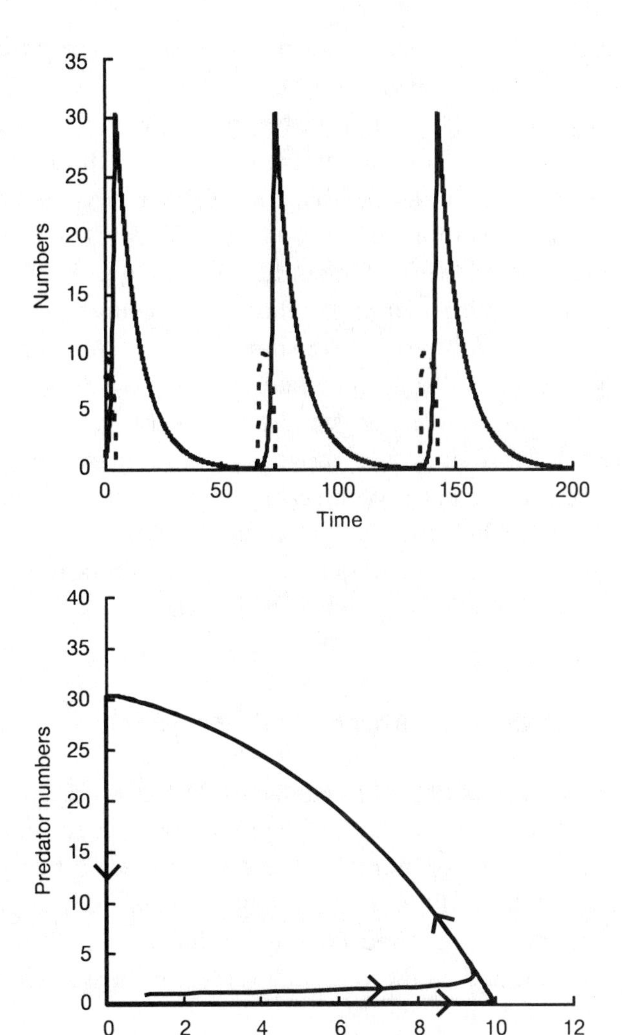

FIGURE 8.14. Dynamics against time (top) and in the phase plane (bottom) for the predator–prey model with a type II functional response, where the equilibrium is unstable. In the plot of numbers against time, the prey is the solid line; the predator is the dashed line. Note how close solutions come to the axes.

location of the predator isocline relative to the 'hump' in the prey isocline. The prey 'hump' moves to the right if the carrying capacity is increased, *so an increase in the carrying capacity of the prey is a destabilizing influence*. This may at first seem counterintuitive, and this concept was originally named the *paradox of enrichment* by Rosenzweig (1971).

Changing the parameter d in the model would also change the location of the 'hump' in the prey isocline. To understand the biological meaning of d, note that if $d = 0$, we recover the original type I functional response. Thus, a small value of d corresponds to a predator that is very efficient at consuming prey, a predator that does not require a large handling time (relative to the rate at which prey are captured) to consume the prey. Also notice that reducing d is a stabilizing influence.

Notice that the parameters a and b do not enter at all into stability.

The other way to change stability in the model would be to alter parameters that would move the location of the predator isocline. Increasing k, the death rate of the predator, would move the predator isocline to the right, and would be a stabilizing influence. Increasing c, the 'efficiency' of converting captured prey into predators, would be a destabilizing factor.

8.6 Further laboratory study of predation

In a very clever study, Luckinbill (1973) followed up Gause's (1934) work with *P. aurelia* and its predator *D. nasutum* with a further series of experiments with the same species that provide a striking qualitative fit to the theory we have just developed. Under the standard controlled conditions that Luckinbill used, the predator consumed all the prey in a few hours, just as Gause had observed.

Luckinbill then altered experimental conditions in two ways that corresponded to the stability-changing ideas we have just discussed. He first tried adding methyl cellulose to the experimental media, which increased the viscosity of the medium and thereby reduced the swimming speed of both organisms. This reduced the overall predation rate of *Didinium*. However, the handling time – the time to actually consume an individual captured prey – remained the same. Thus, the effect on the parameters in the model was to reduce d, which would be a stabilizing influence. Indeed, Luckinbill found that the prey population persisted much longer, although eventually all the prey were consumed. Although this change in conditions was stabilizing, as indicated by the longer persistence time, coexistence was still impossible.

Luckinbill then made one more change in experimental conditions – reducing the food supply for the prey by reducing the density of bacteria (the food for *Paramecium*) in the medium. With this further, stabilizing, change in the experimental conditions, long-term persistence was found. Thus, at least for a simple laboratory system, there is very good *qualitative* agreement between theory and experiment.

8.7 Metapopulation models

We have already seen how a 'patch' structure led to persistence in the experiments performed by Huffaker. We now discuss a simple model capturing some of the aspects of this experiment. As with competition, another approach to predation is to consider the effects of spatial structure in a patch model. We will be looking at a system where at any one location in space the predator–prey interaction is always unstable, and asking whether in a system of patches the 'metapopulation' structure allows predator and prey to coexist. As in our earlier metapopulation models, we ignore any explicit spatial arrangement of the patches we observe. We also look at the fraction of patches in each state, rather than the absolute numbers.

Think of how the assumptions of the model relate to the experiment.

We begin with the possible transitions at one location in space, given in Figure 8.15. In the simplest model each patch is assumed to go through the following sequence: unoccupied, colonization by prey, colonization by predator, extinction of both species, and return to the unoccupied state.

Colonization of empty patches by prey is assumed to be proportional to the fraction of patches occupied only by the prey. Similarly, colonization by the predator of patches occupied by the prey is assumed proportional to the fraction of patches occupied by the predator. Thus the dynamics of the prey are given by the colonization rate of empty patches by the prey minus colonization

Also consider a model where the prey emigrate from patches occupied by the predator.

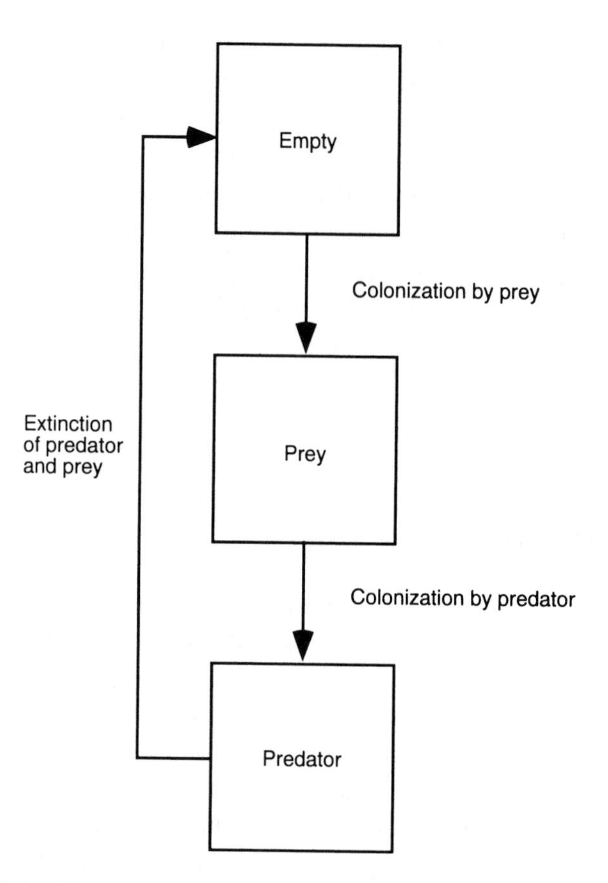

FIGURE 8.15. Transitions among different 'patch' states in a simple spatially structured predator–prey model. A 'patch' labeled 'predator' is one with both predator and prey, where the predator consumes the prey. Here, predators are assumed to be unable to survive in the absence of prey.

of prey patches by the predator;

$$\frac{dp_b}{dt} = m_b p_b(1 - p_b - p_p) - m_p p_b p_p, \tag{8.40}$$

where p_b is the fraction of patches occupied by the prey, m_b is the colonization rate of the prey, and p_p is the fraction of patches occupied by the predator.

The dynamics of the predator are given by the rate of colonization of prey patches minus the extinction rate of patches with the predator. The equation for the dynamics of the predator is

Once again, we assume that predator patches become empty at a fixed rate, recognizing that predator extinction after a fixed time might be more realistic, but more difficult to analyze.

$$\frac{dp_p}{dt} = m_p p_h p_p - e p_p \tag{8.41}$$

where e is the rate at which patches occupied by the predator become empty.

In the problems, you are asked to show that this model invariably leads to a stable equilibrium. Thus, we have shown how a metapopulation structure can lead to persistence of predator and prey that could not coexist within a single habitat. More sophisticated models are needed, however, to produce the oscillatory (rather than equilibrium) dynamics observed by Huffaker (1958).

Think of changes in assumptions that might be destabilizing and lead to oscillations, rather than an equilibrium, in a metapopulation predator–prey model

8.8 Predation in natural systems

How can we use the approach we have developed here to understand natural predator–prey systems? Tanner (1975) attempted to use simple predator–prey models to understand the dynamics of eight pairs of predator and prey. He found rough qualitative agreement between the results from the simple models and whether or not predator–prey cycles were observed in the natural systems.

One of the difficulties of relating the theory we have developed here to interactions between predator and prey in natural systems is finding a relatively contained (in space) system where there is a single predator feeding on a single prey. In most natural systems, predators have alternate prey, and the spatial extent is important, as suggested by our patch model and Huffaker's (1958) experiment. A way to limit the role of spatial extent is to look at an island, which may also serve to limit the number of alternate prey.

The predator–prey interaction between wolves and moose on Isle Royale (located in Lake Superior) has been long studied (Mech, 1966; Taylor, 1984) and can provide insight into the dynamics of predator–prey systems in nature. It is relatively easy to demonstrate that wolves in fact kill moose, but showing that wolves are responsible for regulating the moose population is more difficult; the wolves might only be killing moose that would die anyway. To demonstrate that wolves regulate the moose population, one must show that, as the density of moose increase, the number of deaths from wolf predation also increases. (In all the models, the predation terms include this effect.) The data do not provide a clear-cut answer.

What can essentially be viewed as a field follow-up to Huffaker's study was conducted by Walde et al. (1992), who looked at the dynamics of an herbivorous mite and its predator at two scales: an individual apple tree and the orchard. The study provided strong evidence that spatial structure of the kind studied by Huffaker is important for the persistence of this system. The study also demonstrated that persistence of the predator depended upon alternate prey, again confirming our conclusion that one-predator–one-prey systems are difficult to maintain.

Better evidence for interaction between a single exploiter species and a single victim species comes from the host-parasitoid interactions described in the next chapter.

Problems

1. In this problem we examine the effects of interactions among predators on the stability of a predator–prey system.

   We will modify the basic Lotka–Volterra model without density dependence in the prey, equations (8.3) and (8.4). We will be changing the assumption that the per predator predation rate is independent of the number of predators.

   We assume that predators help each other, as in the case in which many insects can overwhelm the defenses of a plant (which may be mechanical, such as bark on a tree) only if the density of insects is sufficiently high.

   (a) Write down a modified version of the predator–prey model where the predation term (in both the prey and the predator equations) has the per predator predation rate an increasing function of the number of predators. The exact form of the predation rate is not critical. Leave the other two terms alone.

   (b) Draw the prey isocline. (Will it depend on the number of prey?)

(c) Draw the predator isocline; a rough sketch will do. (Which way will it 'tilt' compared to the Lotka–Volterra case?)

(d) Use the graphical technique to determine the signs of the terms in the Jacobian (community) matrix.

(e) Determine the stability of the equilibrium by looking at the sign of the determinant and the trace.

(f) Discuss (verbally) whether this effect of predator interaction makes sense to you – do insects show outbreaks on plants? What role does the fact that the exact form of the predation rate is not critical here have on your drawing conclusions from this model?

2. Here we analyze the metapopulation model, equations (8.40) and (8.41).

   (a) First find the equilibria of the model.

   (b) Then find the stability of the equilibria.

   Does the finding that a stable equilibrium always results correspond to the experimental observations of Huffaker? What might explain the differences?

3. A farmer discovered a pest eating his crops. After spraying with a pesticide, the farmer found that the level of the pest increased! Assume that the pest was being eaten by a predator and that the pesticide affected both the predator and the pest. Also assume that the interaction between the pest and the predator can be described by the continuous time Lotka–Volterra model with density dependence in the prey, equations (8.22) and (8.23). Explain the observed results with this model, by adding to each equation a term representing the effect of the pesticide, an additional source of density-independent mortality for both species.

4. Develop a very simple model where the prey have a refuge by modifying the basic Lotka–Volterra model without density dependence in the prey, equations (8.3) and (8.4).

   (a) For the case where a fraction of the prey are free from predation, first write down the modified model. Then show that the modified model is *mathematically* equivalent to a simple model you have already analyzed: by redefining parameters you recover a model you have seen previously.

(b) For the case where a fixed number of the prey population is free from predation, analyze the model using a phase plane.

(c) Discuss the effects of the refuges on stability, comparing the models to the work of Gause.

Suggestions for further reading

The book edited by Crawley (1992) is a recent, in depth summary of predation (as well as parasitism and disease, which we study next). Gause's original (1934) work and the follow-up by Luckinbill (1973, 1974, 1979) are worth reading. Harrison (1995) summarizes the interplay between theory and Luckinbill's 1973 work.

The paper by Huffaker (1958) is an important classic study. The model for patch dynamics is from Hastings (1977), where more complex models are also considered. See also the models in Gurney and Nisbet (1978), and the greenhouse experiments described by Nachman (1981). Walde et al. (1992) studied a mite predator–prey system in the field analogous to the system looked at by Huffaker in the lab.

9

Host–Parasitoid Interactions

Arthropod predator–prey and host–parasitoid interactions are exceptionally well studied for several reasons. These are systems that can often be studied on a relatively small spatial scale. These also are interactions that can be reasonably thought of as representing tightly coupled systems of two species. Finally, these are systems of great economic importance, because often the hosts are significant pests of crops, and the parasitoids may act as control agents.

A *parasitoid* is a very different kind of predator from a typical hunting predator. Rather than consuming its prey in the normal fashion, a female parasitoid lays its eggs in a particular stage (often larval) of the developing host. The parasitoid eggs then develop within the host, and eventually emerge from the host, typically killing it. It is this requirement that the parasitoid develop within the host that leads to a tight coupling between hosts and parasitoids; often a parasitoid must lay its eggs in only one species of host, to which it is adapted. The typical dynamics of the host–parasitoid interaction are similar to predator–prey dynamics, as illustrated in Figure 9.1.

In this brief chapter, we examine some of the most striking features of host–parasitoid dynamics. Because the interaction is

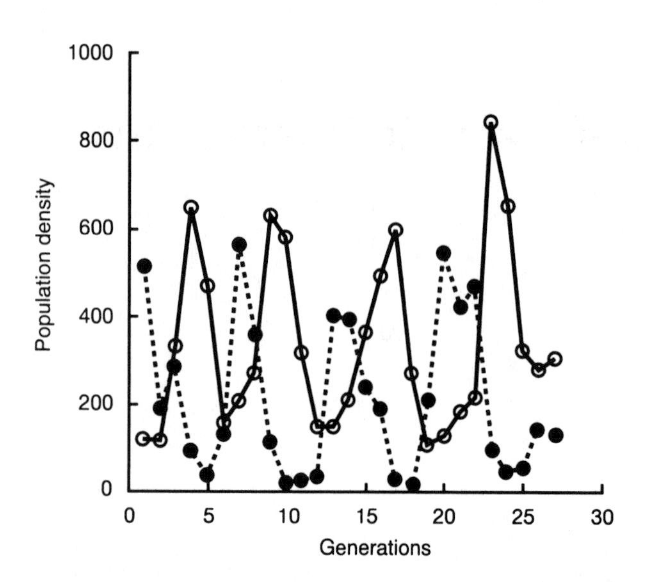

FIGURE 9.1. Dynamics of the host parasitoid interaction between a wasp, *Heterospilus prospidis*, and its host, the bean weevil, *Callosobruchus chinensis* (data are from Utida 1957). —●— Parasitoid; —o— Host

so similar to the predator–prey interaction, an extensive treatment is not needed, and only some of the differences between the two interactions are stressed. It is important to emphasize that the host–parasitoid interaction is not rare – parasitoids represent about 8.5% of described insect species, and estimates of the true fraction of insects that are parasitoids run as high as 25% (Godfray, 1994).

9.1 Nicholson–Bailey model

Once again, our goal is to determine what allows host and parasitoid to persist. The starting point for our discussion is a basic model introduced to ecology in a classic paper by Nicholson and Bailey (1935). Many models of host–parasitoid interactions take this model and its modifications as their starting point.

This is the same Nicholson who performed the classic experiment with blowflies that we examined earlier.

We begin our analysis by looking at a general model for a host–parasitoid system in discrete time. Let N_t be the number of hosts at time t and P_t be the number of parasitoids at time t. We assume that, the host population grows exponentially (geometrically) in

Discrete time is used because hosts often have one or a small fixed number of generations per year, and parasitoids must time their generations to match their hosts.

the absence of the parasitoid at a per capita rate λ per year. Let $f(N_t, P_t)$ be the probability that a host egg laid in year t, with host population size N_t and parasitoid population size P_t, escapes being parasitized. The number of hosts next year is then given by the product of the per capita growth rate of the population, the number of hosts this year, and the probability of escaping parasitism.

We assume that from each parasitized host, c parasitoids emerge the next generation. The number of parasitoids next year is thus given by the product of the number of parasitoids emerging from each parasitized host, the number of hosts this year, and the probability of a host being parasitized. The model we begin with is thus

The model basically assumes that we count at the egg stage in the hosts; otherwise λ would appear in the equation for the parasitoids.

If the probability of not being parasitized is f, what is the probability of being parasitized?

$$N_{t+1} = \lambda N_t f(N_t, P_t) \quad (9.1)$$

$$P_{t+1} = cN_t[1 - f(N_t, P_t)]. \quad (9.2)$$

A first step in using the model will be to determine the functional form, f, which describes the parasitism rate as a function of host and parasitoid density. We make the simplest possible assumptions first. We assume that the parasitoid searches randomly and has an unlimited ability to lay eggs. The description of the distribution obtained by placing objects (parasitoid eggs) randomly into boxes (hosts) is the Poisson distribution. The probability of not being parasitized is just the zero term in the Poisson distribution, which is

$$f = e^{-aP_t}, \quad (9.3)$$

where a is a parameter (with units of 1 over parasitoid numbers) that measures the efficiency of the parasitoid. With this assumption our model becomes the classical Nicholson–Bailey model

$$N_{t+1} = \lambda N_t e^{-aP_t} \quad (9.4)$$

$$P_{t+1} = cN_t[1 - e^{-aP_t}]. \quad (9.5)$$

We make assumptions analogous to the simplest Lotka–Volterra model, and we have studied the effects of delays in one-species models; can you conjecture whether the nontrivial equilibrium of the Nicholson–Bailey model will be stable or unstable?

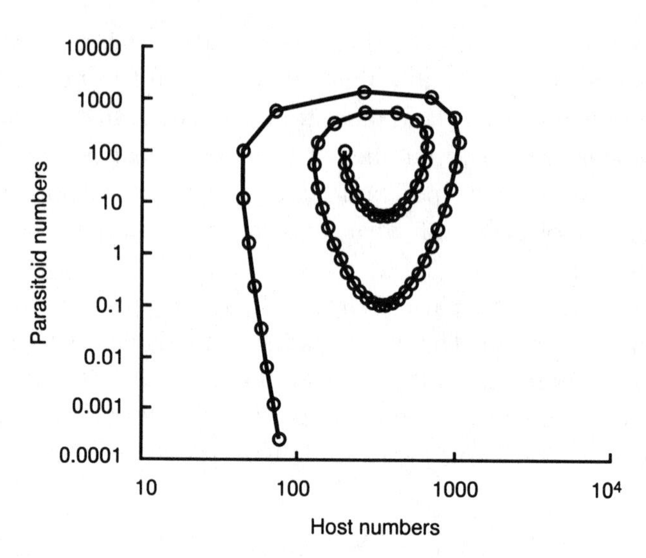

FIGURE 9.2. Dynamics of the Nicholson–Bailey model with $\lambda = 1.1$, $a = 0.001$, and $c = 3$. Note the log-log scale, and that when the simulation was stopped the parasitoid population size had dropped approximately seven orders of magnitude. The dynamics are clearly unstable.

First find P from the equation for N_{t+1} and then find N from the equation for P_{t+1}. Is there another, trivial equilibrium?

We first find the equilibria of the model by setting $N = N_t = N_{t+1}$ and $P = P_t = P_{t+1}$. We obtain the equilibrium

$$N = \frac{\lambda \ln \lambda}{(\lambda - 1)ac} \tag{9.6}$$

$$P = \frac{\ln \lambda}{a}. \tag{9.7}$$

We can immediately see that if this equilibrium is to make biological sense the growth rate of the host in the absence of the predator, λ, must be greater than 1. Also, as the efficiency of the parasitoid goes up – as a increases – the equilibrium level of both species goes down. But what is the biological implication of the equilibrium of this model? Is the equilibrium stable?

We could then proceed with a stability analysis of this model. However, the algebra, is, shall we say, formidable. Thus, rather than go through the steps in the analytical techniques in this chapter, we instead just present the results of stability analyses, and illustrate the behavior of the models with numerical solutions.

Part of the reason that the stability analysis is more difficult here is that the condition that a matrix have both eigenvalues less than one (in magnitude) is more complex than the condition that both eigenvalues have a negative real part. A second reason for the complexity of this analysis is simply the presence of exponentials in the Nicholson–Bailey model.

The results of the stability analysis (which we have omitted) of the Nicholson–Bailey model are unequivocal: the nontrivial equilibrium is unstable for all parameter values. To determine what typical behavior is, we look at dynamics numerically. As illustrated in Figure 9.2, not only is the equilibrium unstable, but the numerical values of the parasitoid levels soon become so low that the only reasonable conclusion is that the model predicts extinction of the parasitoid.

Is it really enough to present results for one set of parameter values?

We are now left with a quandary. Experimental host–parasitoid systems, and natural systems such as the interaction between red scale (*Aonidiella aurantii*) and its parasitoid *Aphytis melinus*, show that long-term coexistence of host and parasitoid is possible. Thus, important stabilizing biological features must be missing from our simple model.

Before reading on, think of stabilizing features of a host–parasitoid interaction that could be included.

9.2 Simple stabilizing features

There have been numerous suggestions of potential stabilizing mechanisms for host–parasitoid systems. We begin by listing a few of the mechanisms that might stabilize the host–parasitoid interaction.

- Density dependence in the host species (Beddington et al., 1975).
- Interference among parasitoids (Beddington et al., 1978). If female parasitoids avoid each other, or avoid laying eggs where others have laid eggs, then the rate of parasitism rises more slowly with the number of parasitoids than assumed in the Nicholson–Bailey model.
- A refuge for the host (Hassell and May, 1974). These refuges can arise in many ways, some of them somewhat counterintuitive. For example, if the host is distributed nonrandomly in space, and the parasitoid preferentially attacks hosts in 'patches' where the hosts are more numerous, then hosts in patches where the hosts are less numerous are essentially in a refuge.

All these concepts can be examined by looking at modifications of the basic Nicholson–Bailey model. We can include density dependence in the prey by the simple modification

Can you think of other ways to include density dependence?

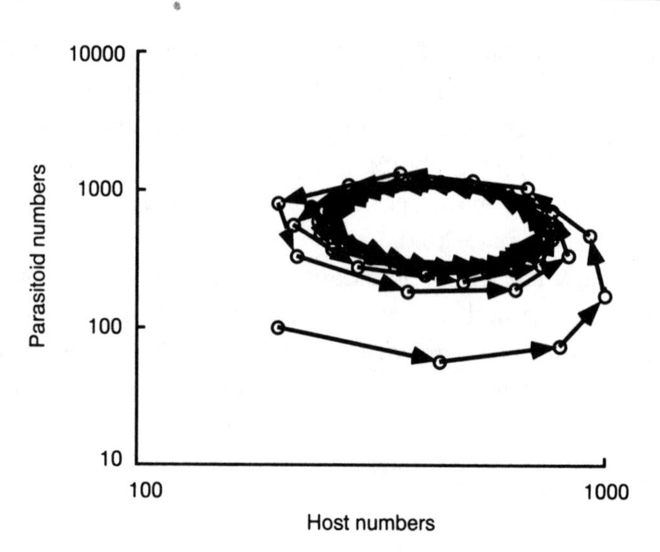

FIGURE 9.3. Dynamics of the Nicholson–Bailey model modified to include density dependence, with $\lambda = 1.1$, $a = 0.001$, $c = 3$, and $K = 1000$. Here host and parasitoid coexist in a stable cycle.

$$N_{t+1} = \lambda N_t e^{[(1-N_t/K)-aP_t]} \tag{9.8}$$
$$P_{t+1} = cN_t[1 - e^{-aP_t}]. \tag{9.9}$$

The stability of this model can again be determined analytically, but we do not do so here. We note that with this modification there is a wide range of reasonable parameters for which the model allows the host and parasitoid to persist, as shown in Figures 9.3 and 9.4. Depending on the parameters, the persistence can either be cyclic (Figure 9.3) or approach a stable equilibrium (Figure 9.4). Thus, the model can lead to the cyclic coexistence illustrated for the laboratory population in Figure 9.1.

From the figures one can see that the parasitoid still does have an effect on the density of the host – the equilibrium (or the average population level for the cyclic persistence) is below the population level the host would reach in the absence of the parasitoid.

From your experience with predator–prey interactions, what might some destabilizing influences?

Similar numerical analyses show that the other two stabilizing features we have listed also can produce stable equilibria.

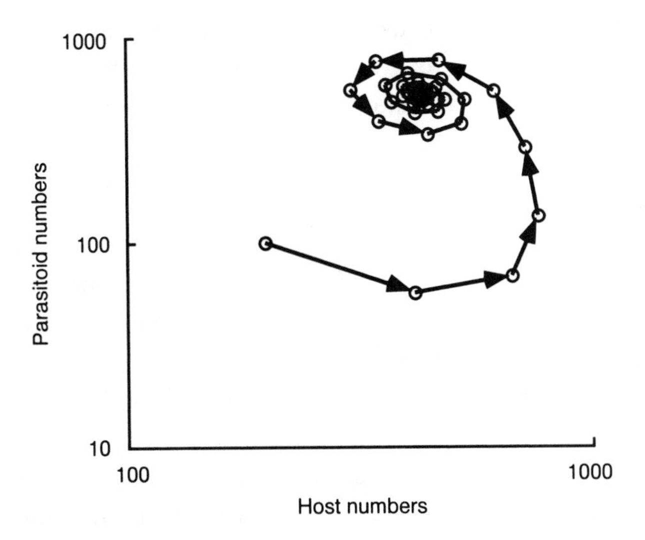

FIGURE 9.4. Dynamics of the Nicholson–Bailey model modified to include density dependence, with $\lambda = 1.1$, $a = 0.001$, $c = 3$, and $K = 750$. Here host and parasitoid coexist at a stable equilibrium.

9.3 What stabilizing features operate in nature?

Models can be used to determine potential stabilizing influences that allow host-parasitoid systems to persist, but cannot be used to determine why any particular pair of species actually do persist in the field. Particularly for host–parasitoid interactions, the unequivocal demonstration that a particular stabilizing force is operating in a field system has been a very difficult task.

One of the best long-term studies attempting to understand what allows host and parasitoid to persist at a stable equilibrium has been that of red scale (*Aonidiella aurantii*) and its parasitoid *Aphytis melinus*, conducted by Murdoch and his colleagues (summarized in Murdoch, 1994). A large number of potential stabilizing mechanisms have been proposed, including many beyond what we have discussed here, but careful experimental examination of all these mechanisms has failed to show that any single one of them is in fact responsible for producing the apparently stable equilibrium observed in the field.

This should not be a cause for pessimism, but should merely indicate that future efforts are needed to unravel the problem of population regulation in field systems.

Problems

1. For the model with density dependence, equations (9.8) and (9.9), would you say that the host population is self-regulating, or regulated by the parasitoid? Justify your answer, and discuss how you would answer this question for a field system.
2. Host–parasitoid systems are used to look at *biological control*, where a parasitoid is introduced to reduce the host population of a pest on a crop. How would the models developed here help to choose a parasitoid to use for biological control, and what qualities should be looked for? What are some potential pitfalls?
3. By numerically solving the model with density dependence, equations (9.8) and (9.9), for the values of a, c, and λ used in the chapter, but varying K, determine the effect of density dependence on stability.

Suggestions for further reading

A summary of host–parasitoid models is contained in Hassell's 1978 book *The Dynamics of Arthropod Predator–Prey Systems*. A recent, extremely comprehensive review of parasitoid biology is given in the book *Parasitoids: Behavioral and Evolutionary Ecology* by Godfray (1994). Murdoch (1990) discusses the relationship of host–parasitoid and predator–prey models to the practice of biological control.

10
Diseases and Pathogens

All of the interactions we have discussed thus far have been between species of approximately the same size. Yet, as ecologists have recently begun to recognize, diseases, pathogens, and parasites may play an extraordinarily important role in regulating populations. One example is shown in Figure 10.1, where both the numbers of larch bud moths and the fraction of moths infected with a particular virus are shown. Once again, we will address the fundamental question of population ecology – can diseases act as the agent preventing runaway exponential population growth? We are also interested in determining what kinds of dynamics are likely to result from the interaction between a species and its pathogens and parasites. In this chapter, we will focus primarily on pathogens (typically bacteria or viruses) where we say that an individual either has a disease or does not. The other possibility is to look at macroparasites, such as tapeworms, where any host individual will carry only a small number of parasites, and differing numbers of parasites have different effects on the state of the host individual.

Can you begin to see a parallel with the metapopulation concepts introduced earlier?

Before looking at the more difficult question of the long-term dynamics of hosts and diseases, we will consider the dynamics of epidemics (Figure 10.2). This question is simpler, because, as il-

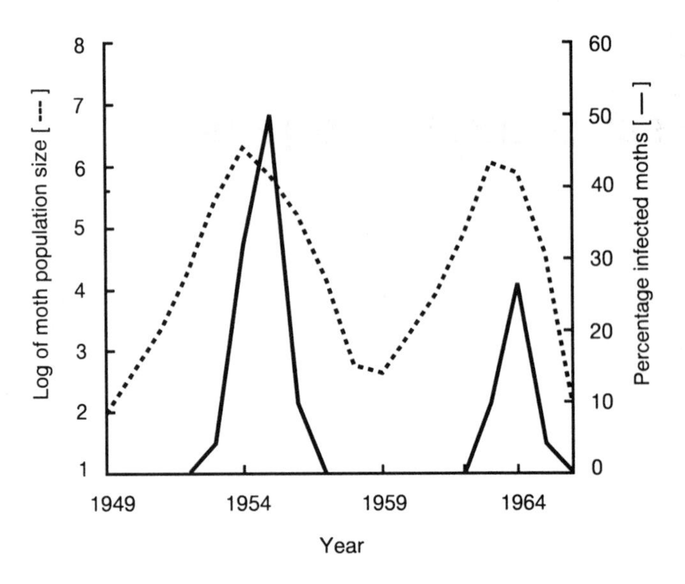

FIGURE 10.1. Dynamics in the Alps of larch bud moth, *Zeiraphera diniana*, and the associated granulosis virus (data are from Auer 1968). Note that the disease apparently acts as a limiting factor to population growth. This, and other examples, are discussed by Anderson and May (1981).

lustrated in the figure, the time course of the epidemic is typically short, very short relative to the normal lifetime of the host. Thus, we can look at a situation where we ignore population dynamics in the host. These are very important problems for public health reasons. Also, before universal vaccination, there were good long-term records of incidence of many childhood diseases (measles, mumps, rubella, chicken pox) in many countries, so extending these ideas to look at diseases over longer time scales provides one of the best ways to determine the usefulness of theories in understanding complex population dynamic processes (Figure 10.3).

10.1 Epidemic models

A basic epidemic model can be used to understand the dynamics of an epidemic like the one illustrated in Figure 10.1. There are two major questions that we will strive to answer.

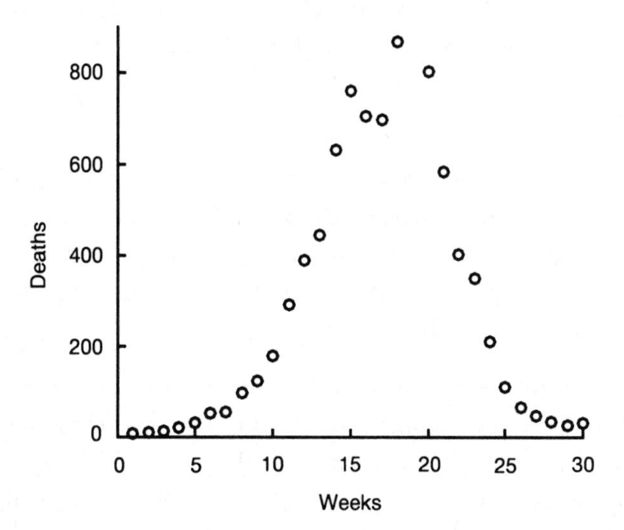

FIGURE 10.2. Deaths from plague in Bombay between December 17, 1905 and July 21, 1906 (data is from Kermack and McKendrick 1927). Note how the disease dies out without killing all members of the host population; the population of Bombay in 1905 was much greater than the total number of deaths indicated in the graph.

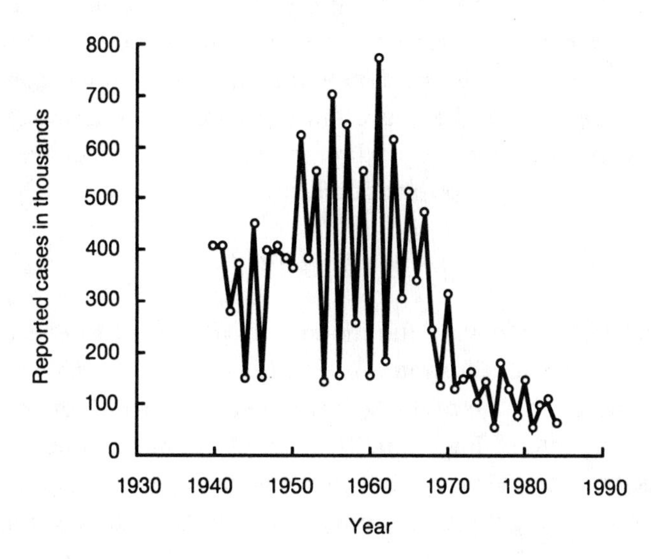

FIGURE 10.3. Reported cases of measles in England and Wales, showing apparently complex dynamical behavior. The decline in the 1970s is the result of vaccination (data from Anderson and May, 1991).

- The epidemic clearly dies out before all the hosts die. Are those individuals who do not get the disease immune, or is there a population dynamic reason for the epidemic to die out?

- Under what conditions will there in fact be an epidemic? What properties of the disease and the population will cause the numbers of individuals with a disease to increase (eventually rapidly) from a presumably small initial inoculum (set of individuals with the disease)?

Again, compare the development here to the metapopulation ideas introduced earlier.

We can answer these questions using a relatively simple model. We will separate the population into three classes, determined by the state of the individuals relative to the disease. Those individuals who do not have the disease, and can potentially get the disease, are called *susceptible*. Individuals who can infect others are called *infective*. Individuals who have either died or recovered, and who can no longer infect others, are called *removed*. Possible transitions among these classes are illustrated in Figure 10.4. In this simplest model, we ignore the delay between when an individual is infected and when the individual becomes infectious.

Into which class should individuals with natural immunity be put?

When using this approach to look at an epidemic, we make the further simplifying assumption that the population is constant. Thus, letting S, I, and R represent the numbers of individuals in the susceptible, infective, and removed classes, respectively, we assume that the total population size

The constant population size assumption makes sense if the time scale of an epidemic is short relative to the lifetime of the host individuals.

$$N = S + I + R \tag{10.1}$$

is a constant. Specifying the model now consists of determining the transition rates illustrated in Figure 10.4.

The rate of change of the susceptible class is given by the negative of the rate of infection. The rate of change of the infective class is given by the rate of infection minus the removal rate, while the rate of change of the removed class is given by the removal rate.

We make the simplest possible assumptions concerning the transition rates. We assume that the rate of infection is βSI, proportional to the product of the number of infectives and susceptibles.

This is like the assumption made in the simplest predator–prey models.

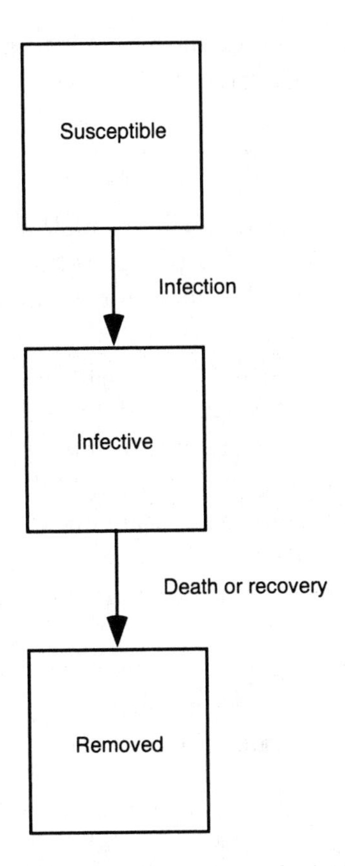

FIGURE 10.4. Box diagram giving the different classes of individuals, and transitions, in a susceptible–infective–removed (SIR) model.

The parameter β, the *contact rate*, depends on both properties of the disease agent and the population being infected.

The only other rate we need to consider is the removal rate. Here we make the simplest assumption of a constant rate of removal, γ.

Thus, as we have seen in similar cases earlier, the mean time an individual has the disease will be $\int_0^\infty e^{-\gamma t} = 1/\gamma$. Thus we can estimate the parameter γ from the mean time that an individual has the disease.

Putting together the transition rates we get the model

$$\frac{dS}{dt} = -\beta SI \tag{10.2}$$

$$\frac{dI}{dt} = \beta SI - \gamma I \tag{10.3}$$

$$\frac{dR}{dt} = \gamma I. \tag{10.4}$$

Why do we choose to eliminate R, not S or I?

Our first step in analyzing this model is to notice that if N is a constant we can use equation (10.1) to write

$$R = N - S - I \tag{10.5}$$

and focus only on the two equations (10.2) and (10.3). Thus, we can understand the dynamics of an epidemic using the phase plane techniques we used to look at competition and predator–prey models.

As a first step, let us look for equilibria. Setting $\frac{dS}{dt} = 0$ and $\frac{dI}{dt} = 0$, we see that I is zero, but S is free to vary. This means that if there are no infectives, the number of susceptibles and infectives remains constant. Also, there is no equilibrium with a non-zero number of infectives. The biological conclusion is that because we have ignored population dynamics in the host, there is no possibility of an *endemic* disease, a disease that is maintained at a steady level in the host.

Divide the right-hand side of equation (10.3) by I.

What can we understand about the dynamics of the disease? An epidemic requires that the number of infectives increase, that $dI/dt > 0$, which is equivalent to

$$\beta S - \gamma > 0. \tag{10.6}$$

We rearrange this condition as

$$R_0 = \frac{\beta S}{\gamma} > 1, \tag{10.7}$$

Often, R_0 is mistakenly called the reproductive rate. However, it is not a rate, which would have units of per time, but a number that has no units. In accord with what we have noted earlier (Box 4.2), parameter combinations controlling dynamics must have no units.

where we call R_0 the *reproductive number* for the disease. This parameter is the key for understanding the dynamics of the disease. We see that βS is the rate at which an infective causes new infections, and, as we have noted, $1/\gamma$ is the mean time an individual is infective. Thus, R_0 is the mean number of new infections caused by a single infective individual. The incidence of the disease will increase if R_0 is greater than 1 and decline if it is less than 1.

We can use this last observation to see that there is a minimum size population in which there can be an epidemic. The maximum number of susceptibles is the total population size, N. So, to have an epidemic, we must have

$$\frac{\beta N}{\gamma} > 1, \tag{10.8}$$

or, rearranging,

$$N > \frac{\gamma}{\beta}. \tag{10.9}$$

We see that there is a minimum population size necessary for an epidemic to occur. There is a large body of evidence corroborating this theoretical prediction: for example, outbreaks of childhood diseases typically occur in cities above certain threshold sizes (Anderson and May, 1991).

We now turn to the other question we have asked – will an epidemic end before all susceptible individuals become infected? To answer this question, we use phase plane methods, but we must use methods that are analytical, rather than graphical. What we will do is compute how the number of infectives change as the number of susceptibles change in an SIR model. We then ask what the number of susceptibles is when the number of infectives is zero. When an epidemic begins, the number of infectives is essentially zero; when an epidemic ends, the same is true. At the end of the epidemic, is the number of susceptibles zero as well? We will show that the answer is no.

From the basic SIR model, we compute

$$\frac{dI}{dS} = -1 + \frac{\gamma}{\beta S} \tag{10.10}$$

by dividing equation (10.3) by equation (10.2). Writing this equation as

$$dI = \left[-1 + \frac{\gamma}{\beta S}\right] dS, \tag{10.11}$$

we can integrate to find that

$$I = \frac{\gamma}{\beta} \ln S - S + C, \tag{10.12}$$

where C is a constant of integration. To find C, we use our knowledge that at the beginning of an epidemic, we essentially have

$I = 0$ and $S = N$. Substituting this into equation (10.12), we get

$$0 = \frac{\gamma}{\beta} \ln N - N + C. \tag{10.13}$$

We solve this last equation for C, finding

$$C = -\frac{\gamma}{\beta} \ln N + N. \tag{10.14}$$

Finally, we get a general expression relating the number of infectives to the number of susceptibles, during the course of an epidemic, by substituting the value for the constant C into equation (10.12), yielding

$$I = \frac{\gamma}{\beta} \ln S - \frac{\gamma}{\beta} \ln N - S + N. \tag{10.15}$$

Finally, we ask at what levels of susceptibles will the numbers of infective be zero. So, we set $I = 0$ in equation (10.15):

$$0 = \frac{\gamma}{\beta} \ln S - \frac{\gamma}{\beta} \ln N - S + N. \tag{10.16}$$

We cannot easily solve this equation for S, which would give us the number of susceptibles at the beginning and end of an epidemic, but we can find out enough information about the solutions to show that at the end of an epidemic the number of susceptibles must be greater than zero. We see that the right-hand side of equation (10.16) is zero when $S = N$, and approaches negative infinity as S approaches 0. Thus, the other value of S that solves equation (10.16) must be much greater than zero. We have demonstrated the remarkable result first elucidated by Kermack and McKendrick (1927) – *an epidemic ends long before all the susceptibles get the disease.*

The natural logarithm of x approaches negative infinity as x approaches 0.

More information about the size of an epidemic can be obtained by solving the equation (10.16) numerically. One can show that the fraction of susceptibles at the end of the disease relative to the total population size, S/N for the smaller value of S solving (10.16) depends only on the ratio $(\gamma/\beta)/N$. The solution to (10.16) is plotted this way in Figure 10.5. From the figure we see confirmation of our two major theoretical results. Even for N slightly above the threshold, an epidemic occurs; and the number of susceptibles at the end of the epidemic is not zero.

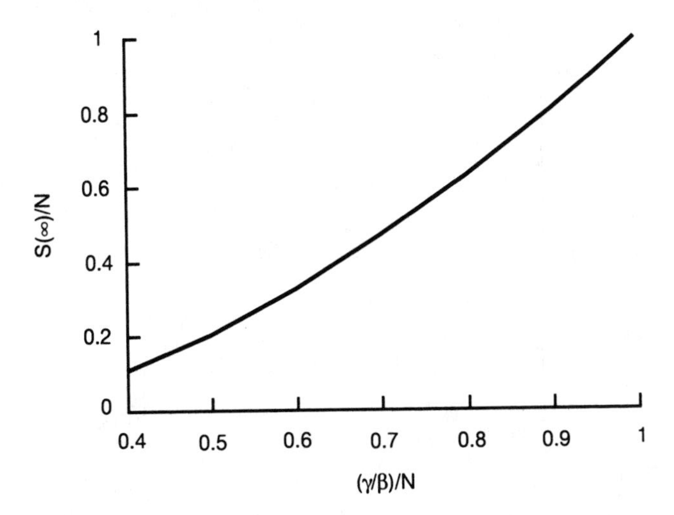

FIGURE 10.5. Plot illustrating the total size of an epidemic. The graph plots the number of susceptibles at the end of an epidemic when t is infinity, $S(\infty)$, divided by population size, N, as a function of how far above the threshold the population size is. When $(\gamma/\beta)/N = 1$, the population size is right at the threshold level. When this ratio drops below 1, an epidemic can occur. The graph shows two important features of epidemics. Even if N is slightly above the level needed to produce an epidemic, the ratio on the horizontal axis is slightly less than 1, a significant number of individuals get the disease. Also, the final number of susceptibles is not 0: the disease dies out before all individuals become diseased.

The full SIR model can be solved either numerically, or approximately analytically, using techniques beyond what we discuss here, producing predictions for the time course of diseases. Doing this, and then picking the best fit to the number of deaths from an epidemic of the plague in Bombay, Kermack and McKendrick produced the astounding fit between model and data shown in Figure 10.6. The SIR model and its extensions have formed the basis for understanding the dynamics of many human diseases (Anderson and May, 1991).

10.2 Can diseases regulate population growth?

We will discuss, using just the concept of reproductive number, whether and how diseases can regulate populations. A detailed answer to this question requires models somewhat more sophis-

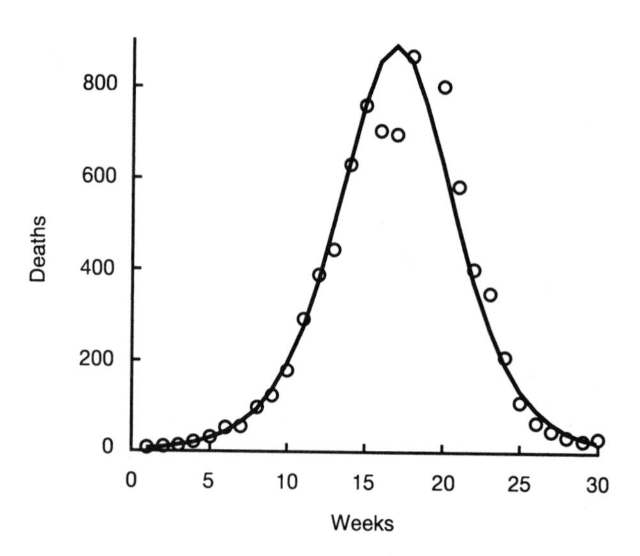

FIGURE 10.6. Deaths per week from plague (open dots) during the 1905–1906 Bombay epidemic, together with fitted SIR model (solid line) (data from Kermack and McKendrick 1927). This is a replot of the data from Figure 10.2 together with a plot of an approximate formula describing disease dynamics according to the SIR model.

Unlike the models for epidemics, here we would have to look at population dynamics in the host.

ticated than we use in this text, with more than two differential equations (see Anderson and May, 1980, 1981).

In the simplest case, think of a population that grows exponentially in the absence of the disease. The disease could reduce the population growth rate by either reducing the birth rate or increasing the death rate of the population. The disease will regulate the population of the host if, in a population in which all the individuals have the disease, the population growth rate would be negative. Otherwise, the disease cannot regulate the population. In the simplest case, when the disease regulates the population the outcome is a stable equilibrium at which the reproductive number of the disease is one.

The situation for the larch bud moth depicted in Figure 10.1 is actually more complex, because the viruses can survive outside the host. However, a simple way to look at the dynamics depicted in the figure is that whenever (the two times) the population size of the larch bud moth rises above a threshold level, then the prevalence of the disease starts to increase. This threshold population level would essentially be the level at which the reproductive number of the disease was one. In this more complex case, the presence of the free-living stage essentially introduces a delay that leads to cycles.

Problems

1. One way that models of epidemics can be used is to understand vaccination strategies. For example, smallpox was eliminated from humans not because every individual in the world was vaccinated, but because a high enough percentage of individuals were vaccinated throughout the world.
   (a) Beginning with the formula (10.9) from the threshold theorem, determine a formula describing the fraction of a population that needs to be vaccinated to make a disease die out.
   (b) The vaccination fraction needed to eliminate smallpox (70%–80%) is lower than that needed to eliminate polio (82%–87%) (Anderson and May, 1991), which in fact has not been eradicated. What does that tell you about differences between the diseases?
2. This is an open-ended question. How would you determine that disease *is* regulating a population, rather than just determining that a disease *could* regulate a population?

Suggestions for further reading

Anderson and May's (1991) book, *Infectious Diseases of Humans*, is an in-depth review of both disease models and disease dynamics in humans. The important issue of the threat of disease for endangered species is reviewed by McCallum and Dobson (1995). Burdon (1987) reviews many aspects of diseases in plant populations.

The regulation of natural populations by diseases also requires looking at evolutionary aspects, because the numerous generations of pathogens in a relatively short time allow natural selection to proceed rapidly. Dwyer et al. (1990) report a detailed study illustrating these concepts in the context of the control of rabbits in Australia by the myxoma virus.

Coda

What regulates populations?

We have answered the question, what *can* regulate populations, but we have not answered the question, what *does* regulate populations. We have seen that populations can be regulated by competitors, predators or parasitoids, diseases, or limited food supplies or cannibalism. This does not tell us, however, what is actually regulating any population in nature. Armed with theories describing potential regulating mechanisms, there are two different conceptual ways of trying to determine what regulates a population in nature.

One could compare the predictions of a model, say one incorporating predation, to the dynamics observed in a field system. For example, predation models make specific predictions about the period of oscillations. These can be compared to estimates of the period of oscillations obtained from time series of the abundances of natural populations. Comparing observed dynamics to predicted model dynamics can clearly reject a proposed explanation of regulation, but cannot prove one.

An alternate approach would be experimental. We have already seen this approach outlined – removal of a putative competitor should produce increases in the density of a competing species. Similarly, augmentation of the food supply for competing species should lead to increases in population density.

The experimental approach, too, has its limitations. Some potential regulating mechanisms, for example, diseases, may be very difficult to manipulate. It may be difficult to manipulate other potential regulating factors on an appropriate temporal or spatial scale (long-lived organisms such as trees or tortoises pose special problems). In cases where experiments can be performed, the theory we have developed in this book is important in suggesting what experiments to perform.

Where do we go from here?

Current research in population biology builds upon the material developed in this book. Syntheses of evolutionary and ecological approaches are needed to understand issues ranging from the mechanisms of speciation to response to global change. There is great interest in developing quantitative connections between models and data – using parameter estimation approaches to really understand how well the models can explain population dynamics.

The emphasis in this book, and in population biology historically, has been to focus on one or two species at a time. There is a great challenge in extending the ideas developed here to understand the dynamics of three or more species at a time. New ideas definitely do emerge.

Many questions of applied interest require approaches based on the material in this book. Management of renewable resources, such as fisheries or forests, depends on using ideas from population ecology. The science of conservation biology depends heavily on making predictions based on ideas from population ecology.

Suggestions for further reading

The wide-ranging review by Murdoch (1994) discusses many issues related to population regulation. The book *Population Dynamics: New Approaches and Synthesis*, edited by Cappuccino and Price (1995), also provides a good overview, focusing primarily on insects. You are also now prepared to read current research in population ecology in major journals such as *American Naturalist, Conservation Biology, Ecological Applications, Ecology, Journal of Animal Ecology, Journal of Ecology, Oikos*, or *Theoretical Population Biology*.

Bibliography

Anderson, R. M., and May, R. M. 1980. Infection diseases and population cycles of forest insects. Science **210**:658–661.

Anderson, R. M., and May, R. M. 1981. The population dynamics of microparasites and their vertebrate hosts. Philosophical Transactions of the Royal Society of London **B291**:451–524.

Anderson, R. M., and May, R. M. 1991. *Infectious Diseases of Humans.* Oxford University Press, Oxford.

Andrewartha, H. G., and Birch, L. C. 1954. *The Distribution and Abundance of Animals.* University of Chicago Press, Chicago.

Auer, C. 1971. Some analyses of the quantitative structure in populations and dynamics of larch bud moth 1949–1968, *in Statistical Ecology, Vol. 2.* Patil, G. P., Peilou, E. C., and Walters, W. E., eds., pp. 151–173. Pennsylvania State University Press, University Park, PA.

Ayala, F. J., Gilpin, M. E., and Ehrenfeld, J. G. 1973. Competition between species: Theoretical models and experimental tests. Theoretical Population Biology **4**:331–356.

Beddington, J. R., Free, C. A., and Lawton, J. H. 1975. Dynamics complexity in predator–prey models framed in difference equations. Nature **255**:58–60.

Beddington, J. R., Free, C. A., and Lawton, J. H. 1978. Characteristics of successful natural enemies in models of biological control of insect pests. Nature **273**:513–519.

Begun, D. J., and Aquadro, C. F. 1995. Molecular variation at the *vermilion* locus in geographically diverse populations of *Drosophila melanogaster* and *D. simulans*. Genetics **140**:1019–1032.

Brown, J. H., and Davidson, D. W. 1977. Competition between seed–eating rodents and ants in desert ecosystems. Science **196**:880–882.

Brown, J. H., Davidson, D. W., and Reichman, O. J. 1979. An experimental study of competition between seed–eating rodents and ants. American Zoologist **19**:1129–1143.

Burdon, J. J. 1987. *Diseases and Plant Population Biology*. Cambridge University Press, Cambridge.

Cappuccino, N., and Price, P. W., eds.1995. *Population Dynamics: New Approaches and Synthesis*. Academic Press, San Diego.

Carpenter, S. R., and Kitchell, J. F., eds. 1993. *The Trophic Cascade in Lakes*. Cambridge University Press, Cambridge.

Caswell, H. 1989. *Matrix Population Models : Construction, Analysis, and Interpretation*. Sinauer Associates, Sunderland, Mass.

Charnov, E. L., and Schaffer, W. M. 1973. Life history consequences of natural selection: Cole's result revisited. American Naturalist **107**:791–793.

Cole, L. C. 1954. The population consequences of life history phenomena. Quarterly Review of Biology **29**:103–137.

Congdon, J. D., Dunham, A. E., and van Loben Sels, R. C. 1993. Delayed sexual maturity and demographics of Blanding's turtles (*Emydoidea blandingii*): Implications for conservation and management of long–lived organisms. Conservation Biology **7**:826–833.

Congdon, J. D., Dunham, A. E., and van Lobel Sels, R. C. 1994. Demographics of common snapping turtles (*Chelydra serpentina*): Implications for conservation and management of long–lived organisms. American Zoologist **34**:397–408.

Connell, J. H. 1961. The influence of interspecific competition and other factors on the distribution of the barnacle *Chthamalus stellatus*. Ecology **42**:710–723.

Connell, J. H. 1983. On the prevalence and relative importance of interspecific competition. American Naturalist **122**:661–696.

Connell, J. H., and Sousa, W. P. 1983. On the evidence needed to judge ecological stability or persistence. American Naturalist **121**:789–824.

Costantino, R. F., Cushing, J. M., Dennis, B., and Desharnais, R. A. 1995. Experimentally induced transitions in the dynamic behaviour of insect populations. Nature **375**:227–230.

Crawley, M. J. 1990. The population dynamics of plants. Philosophical Transactions of the Royal Society of London **B330**:125–140.

Crawley, M. J., ed. 1992. *Natural Enemies: The Population Biology of Predators, Parasites, and Diseases*. Blackwell Scientific Publications, Oxford.

Crow, J. F. 1986. *Basic Concepts in Population, Quantitative, and Evolutionary Genetics*. W.H. Freeman, San Francisco.

Davidson, J. 1938. On the growth of the sheep population in Tasmania. Transactions of the Royal Society of South Australia **62**:342–346.

Dawson, P. S. 1970. Linkage and the elimination of deleterious mutant genes from experimental populations. Genetica (Dordrecht) **41**:147–169.

Doak, D., Kareiva, P., and Kleptetka, B. 1994. Modeling population viability for the desert tortoise in the western Mojave desert. Ecological Applications **4**:446–460.

Dwyer, G., Levin, S. A., and Buttel, L. 1990. A simulation model of the population dynamics and evolution of myxomatosis. Ecological Monographs **60**:423–447.

Endler, J. A. 1986. *Natural Selection in the Wild*. Princeton University Press, Princeton.

Evans, F. C., and Smith, F. E. 1952. The intrinsic rate of natural increase for the human louse *Pediculus humanus* L. American Naturalist **86**:299–310.

Falconer, D. S. 1989. *Introduction to Quantitative Genetics, 3rd Ed.*. Longman, London.

Fisher, R. A. 1930. *The Genetical Theory of Natural Selection*. Clarendon Press, Oxford.

Futuyma, D. J. 1986. *Evolutionary Biology*. Sinauer Associates, Sunderland, Mass.

Gause, G. F. 1934. *The Struggle for Existence*. Williams & Wilkins, Baltimore.

Gause, G. F. 1935. *Verifications Experimentales de la Theorie Mathematique de la Lutte pour la Vie*. Hermann, Paris.

Gillespie, J. H. 1991. *The Causes of Molecular Evolution*. Oxford University Press, Oxford.

Gilpin, M. E., and Ayala, F. J. 1973. Global models of growth and competition. Proceedings of the National Academy of Sciences of the U.S.A. **70**:3590–3593.

Gilpin, M., and Hanski, I., eds. 1991. *Metapopulation Dynamics: Empirical and Theoretical Investigations*. Academic Press, London.

Godfray, H. C. J. 1994. *Parasitoids: Behavioral and Evolutionary Ecology*. Princeton University Press, Princeton.

Gurney, W. S. C., Blythe, S. P., and Nisbet, R. M. 1980. Nicholson's blowflies revisited. Nature **287**:17–21.

Gurney, W. S. C., and Nisbet, R. M. 1978. Predator–prey fluctuations in patchy environments. Journal of Animal Ecology **47**:85–102.

Hairston, N. G., Smith, F. E., and Slobodkin, L. B. 1960. Community structure, population control, and competition. American Naturalist **94**:421–425.

Halbach, U. 1979. Introductory remarks: strategies in population research exemplified by rotifer population dynamics. Fortschritte der Zoologie **25**:1–27.

Hamilton, W. D., and May, R. M. 1977. Dispersal in stable habitats. Nature **269**:578–581.

Hanski, I. 1990. Density dependence, regulation and variability in animal populations. Philosophical Transactions of the Royal Society of London **B330**:19–28.

Harper, J. L., and McNaughton, I. H. 1962. The comparative biology of closely related species living in the same area. VII. Interference between individuals in pure and mixed populations of *Papaver* species. New Phytologist **61**:175–188.

Harrison, G. W. 1995. Comparing predator–prey models to Luckinbill's experiment with *Didinium* and *Paramecium*. Ecology **76**:357–374.

Hartl, D. L., and Clark, A. G. 1988. *Principles of Population Genetics, 2nd Ed.*. Sinauer Associates, Sunderland, Mass.

Hassell, M. P. 1978. *The Dynamics of Arthropod Predator–Prey Systems*. Princeton University Press, Princeton, N.J.

Hassell, M. P., Latto, J., and May, R. M. 1989. Seeing the wood for the trees: detecting density dependence from existing life–table studies. Journal of Animal Ecology **58**:283–300.

Hassell, M. P., and May, R. M. 1974. Aggregation in predators and insect parasites and its effect on stability. Journal of Animal Ecology **43**:567–594.

Hastings, A. 1977. Spatial heterogeneity and the stability of predator–prey systems. Theoretical Population Biology **12**:37–48.

Hastings, A. 1980. Disturbance, coexistence, history, and competition for space. Theoretical Population Biology **18**:363–373.

Hastings, A., and Harrison, S. 1994. Metapopulation dynamics and genetics. Annual Review of Ecology and Systematics **25**:167–188.

Hastings, A., Hom, C. L., Ellner, S., Turchin, P., and Godfray, H. C. J. 1993. Chaos in ecology: Is mother nature a strange attractor? Annual Review of Ecology and Systematics **24**:1–33.

Hengeveld, R. 1989. *Dynamics of Biological Invasions*. Chapman & Hall, New York.

Holling, C. S. 1959. Some characteristics of simple types of predation and parasitism. Canadian Entomologist **91**:385–398.

Holling, C. S. 1973. Resilience and stability of ecological systems. Annual Review of Ecology and Systematics **4**:1–23.

Hudson, R. 1965. The spread of the collared dove in Britain and Ireland. British Birds **58**:105–139.

Hudson, R. 1972. Collared doves in Britain and Ireland during 1965–1970. British Birds **65**:139–155.

Huffaker, C. B. 1958. Experimental studies on predation: dispersion factors and predator-prey oscillations. Hilgardia **27**:343–383.

Hutchinson, G. E. 1948. Circular causal systems in ecology. Annals of the New York Academy of Sciences **50**:221–246.

Hutchinson, G. E. 1978. *An Introduction to Population Ecology*. Yale Unviersity Press, New Haven.

Johnson, M. L., and Gaines, M. S. 1990. Evolution of dispersal: theoretical models and empirical tests using birds and mammals. Annual Review of Ecology and Systematics **21**:449–480.

Kermack, W. O., and McKendrick, A. G. 1927. A contribution to the mathematical theory of epidemics. Proceedings of the Royal Society of London **A115**:700–721.

Kettlewell, H. B. D. 1956. A resume of investigations on the evolution of melanism in the Lepidoptera. Proceedings of the Royal Society of London **B145**:297–303.

Kimura, M. 1968. Evolutionary rate at the molecular level. Nature **217**:624–626.

Kimura, M. 1983. *The Neutral Theory of Molecular Evolution*. Cambridge University Press, Cambridge.

Kingsland, S. 1985. *Modeling Nature*. University of Chicago Press, Chicago.

Leslie, P. H. 1945. On the use of matrices in certain population mathematics. Biometrika **33**:183–212.

Leslie, P. H. 1948. Some further notes on the use of matrices in population analysis. Biometrika **35**:213–245.

Leslie, P. H., and Ranson, R. M. 1940. The mortality, fertility and rate of natural increase of the vole (*Microtus agrestis*) as observed in the laboratory. Journal of Animal Ecology **9**:27–52.

Levins, R. 1966. The strategy of model building in ecology. American Scientist **54**:421–431.

Levins, R. 1969. Some demographic and genetic consequences of environmental heterogeneity for biological control. Bulletin of the Entomological Society of America **15**:237–240.

Lewontin, R. C. 1969. The meaning of stability. Brookhaven Symposia in Biology **22**:13–24.

Lewontin, R. C., and Hubby, J. L. 1966. A molecular approach to the study of genic heterozygosity in natural populations. II. Amount of variation and degree of heterozygosity in natural populations of *Drosophila pseudoobscura*. Genetics **54**:595–609.

Lotka, A. J. 1926. *Elements of Physical Biology*. Williams & Wilkins, Baltimore.

Lotka, A. J. 1932. The growth of mixed populations: Two species competing for a commonn food supply. Journal of the Washington Academy of Sciences **22**:461–469.

Luckinbill, L. S. 1973. Coexistence in laboratory populations of *Paramecium aurelia* and *Didinium nasutum*. Ecology **54**:1320–1327.

Luckinbill, L. S. 1974. The effects of space and enrichment on a predator–prey system. Ecology **55**:1142–1147.

Luckinbill, L. S. 1979. Regulation, stability, and diversity in a model experimental microcosm. Ecology **60**:1098–1102.

MacLulick, D. A. 1937. Fluctuations in numbers of the varying hare (*Lepus americanus*). University of Toronto Studies, Biology Series **43**:1–136.

May, R. M. 1974. Biological populations with non–overlapping generations: stable points, stable cycles, and chaos. Science **186**:645–647.

May, R. M. 1975. *Stability and Complexity in Model Ecosystems, 2nd Ed.*. Princeton University Press, Princeton.

May, R. M. 1976. Simple models with very complicated dynamics. Nature **261**:459–467.

Maynard Smith, J. 1982. *Evolution and the Theory of Games.* Cambridge Unviersity Press, Cambridge.

Maynard Smith, J. 1989. *Evolutionary Genetics.* Oxford University Press, Oxford.

McCallum, H., and Dobson, A. 1995. Detecting disease and parasite threats to endangered species and ecosystems. Trends in Ecology & Evolution **10**:190–194.

McKendrick, A. G., and Pai, M. K. 1911. The rate of multiplication of microorganisms: a mathematical study. Proceedings Royal Society of Edinburgh **31**:649–655.

Mech, L. D. 1966. The wolves of Isle Royale. Fauna of National Parks of the United States. Fauna Series **7**:1–210.

Murdoch, W. W. 1990. The relevance of pest–enemy models to biological control, *in Critical Issues in Biological Control.* Mackauer, M., Ehler, L. E., and Roland, J., eds., pp. 1–24. Intercept, Andover, Hants.

Murdoch, W. W. 1994. Population regulation in theory and practice. Ecology **75**:271–287.

Nachman, G. 1981. Temporal and spatial dynamics of an acarine predator–prey system. Journal of Animal Ecology **50**:435–451.

Nee, S., and May, R. M. 1992. Dynamics of metapopulations —- habitat destruction and coempetitive coexistence. Journal of Animal Ecology **61**:37–40.

Nicholson, A. J. 1957. The self–adjustment of populations to change. Cold Spring Harbor Symposia on Quantitative Biology **22**:153–173.

Nicholson, A. J., and Bailey, V. A. 1935. The balance of animal populations. Proceedings of the Zoological Society of London **1**:551–598.

Nisbet, R. M., and Gurney, W. S. C. 1982. *Modelling Fluctuating Populations.* John Wiley & Sons, New York.

Park, T. 1948. Experimental studies of interspecific competition. I. Competition between populations of the flour beetle *Tribolium confusum* Duval and *Tribolium castaneum* Herbst. Ecological Monographs **18**:267–307.

Park, T. 1954. Experimental studies of interspecific competition. I. Temperature, humidity and competition in two species of *Tribolium*. Physiological Zoology **27**:177–238.

Pearl, R., Reed, L. J., and Kish, J. F. 1940. The logistic curve and the census count of 1940. Science **92**:486–488.

Pielou, E. C. 1981. The usefulness of ecological models: a stock taking. Quarterly Review of Biology **56**:17–31.

Polis, G. A., and Winemiller, K. O., eds.1996. *Food Webs*, Chapman & Hall, New York.

Real, L. A., ed.1994. *Ecological Genetics*. Princeton University Press, Princeton.

Ricker, W. 1954. Stock and recruitment. Journal of the Fisheries Research Board of Canada **11**:559–623.

Roff, D. A. 1992. *The Evolution of Life Histories*. Chapman & Hall, New York.

Rosenzweig, M. L. 1971. Paradox of enrichment: Destabilization of exploitation ecosystems in ecological time. Science **171**:385–387.

Roughgarden, J. 1979. *Theory of Population Genetics and Evolutionary Ecology: An Introduction*. MacMillan Publishing Co., New York.

Roughgarden, J., May, R. M., and Levin, S. A., eds.1989. *Perspectives in Ecological Theory*. Princeton University Press, Princeton.

Salceda, V.M. 1977. Carga genetica en siete poblaciones naturales de *Drosophila melanogaster* (Meigen) de diferentes localidades de Mexico. Sobretiro de Agrociencia **28**:47–52.

Schaffer, W. 1974. Selection for optimal life histories: The effects of age structure. Ecology **55**:291–303.

Schaffer, W. M., and Gadgil, M. D. 1975. Selection for optimal life histories in plants, *in Ecology and Evolution of Communities*. Cody, M. L., and Diamond, J. M., eds., pp. 142–157. Harvard University Press, Cambridge, Mass.

Scharloo, W., Hoogmoed, M. S., and Kuile, A. T. 1967. Stabilizing and disruptive selection on a mutant character in *Drosophila*. I.

The phenotypic variance and its components. Genetics **56**:709–726.

Schoener, T. W. 1974. Competition and the form of habitat shift. Theoretical Population Biology **6**:265–307.

Schoener, T. W. 1983. Field experiments on interspecific competition. American Naturalist **122**:240–285.

Shelford, V. E. 1943. The relation of snowy owl migration to the abundance of the collared lemming. Auk **62**:592–594.

Shepherd, J. G., and Cushing, D. H. 1990. Regulation in fish populations: myth or mirage? Philosophical Transactions of the Royal Society of London **B330**:151–164.

Skellam, J. G. 1951. Random dispersal in theoretical populations. Biometrika **38**:196–218.

Slobodkin, L. B. 1961. *Growth and Regulation of Animal Populations.* Holt, Rinehart, and Winston, New York.

Slobodkin, L. B. 1964. Experimental populations of Hydrida. Journal of Animal Ecology **33**:131–148.

Stearns, S. C. 1992. *The Evolution of Life Histories.* Oxford University Press, Oxford.

Tanner, J. T. 1975. The stability and intrinsic growth rate of prey and predator populations. Ecology **56**:855–867.

Taylor, R. J. 1984. *Predation.* Chapman & Hall, New York.

Tilman, D. 1994. Competition and biodiversity in spatially structured habitats. Ecology **75**:2–16.

Utida, S. 1957. Cyclic fluctuations of population density intrinsic to the host–parasite system. Ecology **38**:442–449.

Volterra, V. 1926. Variazioni e fluttuazioni del numero d'individui in specie animali conviventi. Memoirie del Acadamei Lincei **2**:31–113.

Volterra, V. 1931. *Lecons sur la Mathematique de la Lutte pour la Vie.* Marcel Brelot, Paris.

Walde, S. J., Nyrop, J. P., and Hardman, J. M. 1992. Dynamics of *Panonychus ulmi* and *Typhlodromus pyri*: factors contributing to persistence. Experimental & Applied Acarology **14**:261–291.

Wangersky, P. J., and Cunningham, W. J. 1957. Time lag in prey-predator models. Ecology **38**:136–139.

Index

Numbers in boldface indicate where a term is defined.